ELECTRONICS

Explained Simply and Easily

A Quick Guide to Electronic Fundamentals,
Components, Circuits and Applications

Brought to you by

VPACHKAWADE Research Center

www.vpachkawade.com

contact@vpachkawade.com

No part of this book may be reproduced, distributed, or transmitted in any form or by any means, including photocopying, recording, or other electronic or mechanical methods, without the prior written permission of the publisher, except in the case of brief quotations embodied in critical reviews and certain other noncommercial uses permitted by copyright law.

Disclaimer of Liability:

The author and publisher have made every effort to ensure the accuracy of the information in this book but do not guarantee its completeness or suitability for any purpose. This book is provided on an "as-is" basis, and the author and publisher shall not be held liable for any damages or losses arising from its use.

Trademark Acknowledgment:

All trademarks and brand names mentioned in this book are the property of their respective owners.

For permissions, inquiries, or feedback, please contact:

contact@vpachkawade.com

Page intentionally left blank

Contents

What will you learn from this book?

Learners can expect to achieve following learning objectives or outcomes after reading this book.

- Learn about electronic components, their symbols, and functions within circuits

- Gain proficiency in analyzing electronic components and circuits using fundamental laws and techniques

- Learn details on KVL, KCL, diode, diode circuits, transistor and transistor circuits, biasing methods, etc.

- Learn basics and advances of electronic components, semiconductor devices (diode, BJT, MOSFET, op-amp, etc)

- Develop skills in designing and prototyping electronic circuits on breadboards and troubleshooting circuit designs

- Learn skills to test and make measurements in electronics design

- Learn how to use software (CAD) to design, simulate and analyze circuits

- Learn practical electronics - do it yourself practice

Are there requirements/prerequisites for reading this book?

- No specific requirements or prior experience necessary - this beginner-friendly book on Electronics welcomes all interested learners.

Who is this book for?

Below is the description of the intended learners for our book who will find our book content valuable.

- This book is intended for beginners and enthusiasts who want to learn about Electronics.

- Undergraduate/graduate students in engineering, science and technology in all disciplines

- Electronics hobbyists

- Engineers and technicians

- Anyone curious about how electronics work and how to use and apply it to develop practical applications.

- Whether you have a background in electronics or are completely new to the subject, this course will provide a solid foundation and practical knowledge to start working with electronics and undertake exciting projects.

- Professionals willing to revise the concepts in electronic components, semiconductor devices and circuits analysis

Book Description

Are you eager to understand the world of electronics? Whether you're an aspiring engineer, a tech enthusiast, or someone simply curious about how the gadgets you use every day work, our "Electronics - A Quick Guide to Electronic Fundamentals, Circuits and Applications" is your ticket to understanding the fundamental principles that power the modern world.

Book Highlights:

Foundations of Electronics: Dive into the very core of electronics by exploring concepts like voltage, current, resistance, capacitance and power. Understand Ohm's law and how it governs the behavior of electronic components.

Electronic Components: Discover the key building blocks of electronic circuits, including resistors, capacitors, inductors, diodes, and transistors. Learn how to identify, use, and connect these components effectively.

Circuit Analysis: Master the art of analyzing simple electronic circuits. You'll develop the skills to calculate voltages, currents, and power across different components using various circuit analysis techniques. Learn details on KVL and KCL with examples.

Basic Circuit Design: Start designing your circuits! Explore different types of circuits such as voltage dividers, amplifiers, and filters. Get hands-on experience designing and building your own basic circuits.

Semiconductor Devices: Uncover the magic of semiconductors. Learn about diodes and transistors, their types, functions, and applications in modern electronics. Learn details on diode, diode circuits, transistor and transistor circuits, biasing methods, etc.

Practical Applications: Discover how Electronics concepts apply to real-world applications. From simple gadgets to complex systems, you'll see how electronics is everywhere.

Troubleshooting and Repair: Equip yourself with essential troubleshooting skills to identify and fix common electronic problems. Learn how to use multimeters and oscilloscopes effectively.

Safety in Electronics: Safety is paramount when working with electronics. We'll cover best practices for safely handling electronic components and equipment.

This book contains valuable information such as reference materials, component datasheets, circuit diagrams, and additional resources such as links to video lectures on our YouTube channel, VPACHKAWADE Research Center

Link/s to all related YouTube lectures for easy references

https://www.youtube.com/playlist?list=PLWQzGc-sfVeWR-

UJ5xlcJMLVevHI3SA2V

Chapter 1 Introduction

I. Electronics and its importance in today's world

Figure 1-1 Emerging application areas of electronics and the research being carried out at VPACHKAWADE Research center. For details visit, www.vpachkawade.com

Electronics plays a pivotal role in today's world, shaping nearly every aspect of our lives. From the smartphones we carry in our pockets to the advanced machinery powering industries, electronics are the heartbeat of modern society. They have revolutionized communication, enabling instant global connections. Additionally, electronics drive innovations in healthcare, transportation, entertainment, and countless other domains, improving efficiency and quality of life. Moreover, they are at the core of renewable energy solutions, contributing to a more sustainable future. In essence, electronics are the foundation upon which our modern world is built, and their importance continues to grow as technology evolves. Figure 1-1 shows several possible emerging areas of electronics technology and applications.

II. **Modern surface mount electronics components | SMD**

Figure 1-2 showcases an electronics development board, specifically a printed circuit board (PCB). The PCB hosts an array of modern electronic components, each thoughtfully placed on this circuit board.

Within this assembly, you can see a series of surface-mount resistors. Additionally, there's a surface-mount capacitor. There are also two distinct components, possibly LEDs or other similar devices. Furthermore, there's an integrated chip, commonly referred to as an IC. You can identify this IC by referring to the data number on its package, and you can look up its datasheet on Google to learn more about its specific functionality.

Integrated circuit (IC)

capacitor

resistors

Figure 1-2 Modern surface-mount electronic components on a printed circuit board (PCB), with a large IC at the top [60].

Similarly, the codes printed on the bodies of these resistors, as well as other components, provide crucial information about their values. You can consult the manufacturer's datasheets to gain a deeper understanding of these components.

One noticeable feature is a regulator IC present on the board, with an identifiable code on its body. A simple Google search can help you determine the type of component or IC that this represents.

This is an example of what modern electronic development boards look like, featuring miniaturized versions of electronic components and integrated circuits.

III.   Basic electronic components and their symbols

In the world of electronics, there are several fundamental components that serve as the building blocks for all electronic devices. Understanding these components and their symbols is essential for anyone diving into the fascinating realm of electronics. Figure 1-3 shows various electronic components used in modern electronics. Electronic components are fundamental parts used to build electronic circuits and devices. They can be categorized into two main types: passive and active components. Passive Components are basic electronic elements that do not require an external power source to function. They include: resistors, capacitors, inductors, and diodes. Active Components need an external power source to operate and control electrical current. They include: diodes, transistors, integrated circuits such as Op-

Amps, voltage regulators, LED (Light Emitting Diode) etc.

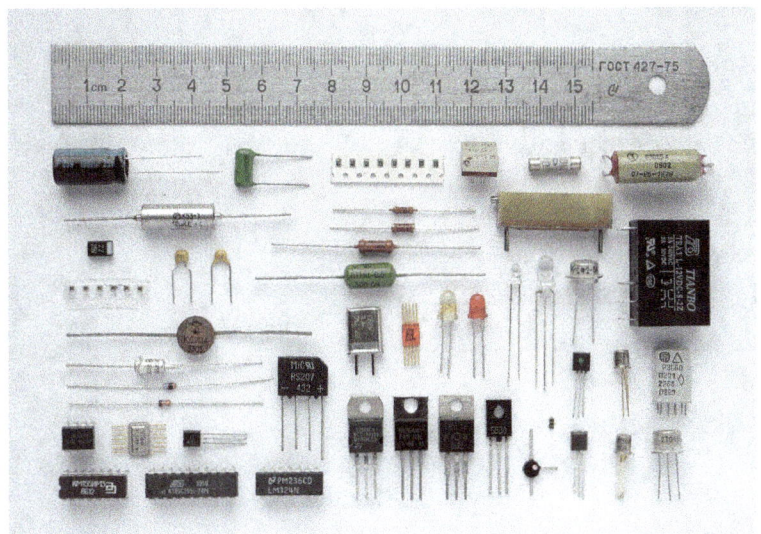

Figure 1-3 Various electronic components [1]

Figure 1-4 shows range of electronic components **symbols** used in common electronics applications. We shall also explore details of each of these components in the following sections.

Resistor (R):

A resistor limits the flow of electric current in a circuit. The symbol for a resistor is the Greek letter omega (Ω), and its value is measured in ohms (Ω). Resistors come in various sizes and shapes, and their job is to resist or reduce the flow of electricity.

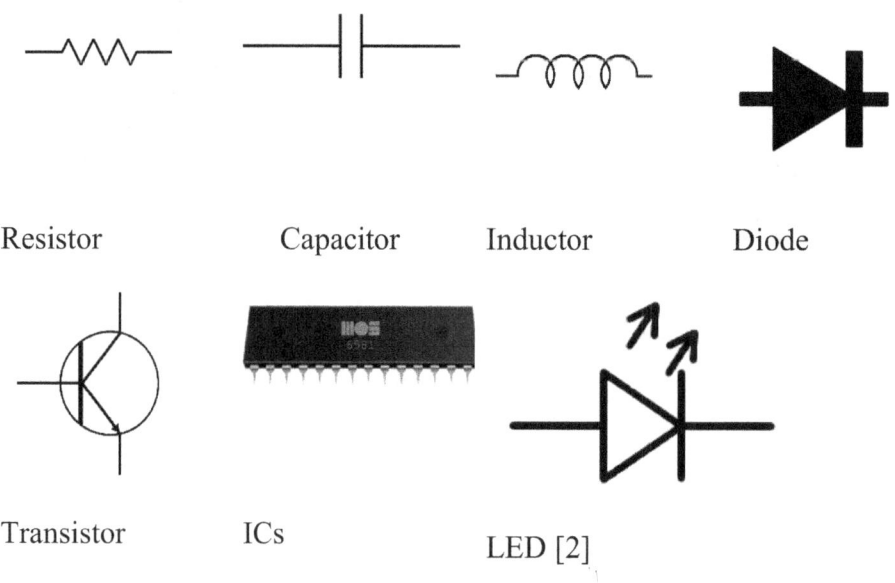

Resistor	Capacitor	Inductor	Diode
Transistor	ICs	LED [2]	

Figure 1-4 Various electronic components widely used in electronic - symbols

Capacitor (C):

Imagine a capacitor as a temporary battery. It stores electrical energy and releases it when needed. Capacitors are used for various purposes, from smoothing out power supplies to timing circuits.

Inductor (L):

An inductor is like a coil of wire. It stores electrical energy in a magnetic field. Inductors resist changes in current flow, making them valuable in filters and energy storage.

Diode (D):

A diode allows current to flow in only one direction. They play a critical role in rectifying AC (alternating current) to DC (direct current) and protecting circuits from reverse voltage.

Transistor (NPN, PNP):

Transistors are like electronic switches that can amplify or control current. There are different types, such as NPN and PNP. They are essential for amplifying signals and building logic gates in digital circuits.

Integrated Circuit (IC):

An integrated circuit is like a tiny city of electronic components packed into one chip. ICs contain multiple electronic components, like transistors and resistors, all working together to perform various functions. They're the brains behind modern electronic devices.

These are the basic electronic components you'll encounter frequently in electronic circuits. Understanding their symbols and functions is important in electronics. Each component has a unique role in shaping the behavior of circuits, making them the foundation of all electronic devices.

Light Emitting diode (LED):

An LED, or Light Emitting Diode, is a tiny but powerful electronic device that shines with vibrant colored light. LEDs work by allowing an electric current to flow through them, which triggers them to glow. They are commonly used in various everyday things like digital displays, indicator lights on appliances, and even in dazzling light shows. LEDs are favored for their energy efficiency, lasting a long time, and their ability to produce bright and colorful light. Whether it's the indicator on your phone or the colorful decorations during the holiday season, LEDs make our world a little brighter and more colorful.

IV.   Overview of electronic circuits

Figure 1-5 Range of electronic components used to build circuits, and other useful functions of electronics hardware

In our exploration of electronics, it's essential to start with an overview of electronic circuits. Imagine circuits as the roadways of the electronic world, guiding electricity to perform specific tasks. These circuits are built from basic components like resistors, capacitors, diodes, and transistors, each with its unique job. We use symbols to represent these components on paper. Power sources, such as batteries, provide the energy needed to make circuits work. There are different types of circuits, some for carrying continuous signals (analog) and others for processing digital information (digital).

Circuit diagrams help us understand how everything is connected. In our journey, we'll explore various circuit configurations, uncover their functions, and see how they power the devices that make our lives easier and more exciting.

Chapter 2   Basic Electronic Components

I.   Resistors

We have learned earlier elementary knowledge about what resistor is and how it works, its relationship with voltage and current through ohm's law. Resistors are widely used in design of electronic circuits and systems, where, they are used for voltage division, current limiting, and signal conditioning.

a. Types of Resistors

a) Fixed Resistors: There are two types here.

- **Carbon Composition Resistors:**

Carbon composition resistors are traditional passive electronic components known for their simplicity and durability (see Figure 2-1). These resistors are constructed using a mixture of finely ground carbon particles and a ceramic or clay binder. This combination forms a resistive element that hinders the flow of electric current. Carbon composition resistors have been widely used in various electronic applications for decades. They are characterized by their ruggedness, high voltage-handling capabilities, and resistance to mechanical stress and vibration. However, they may have limitations in terms of precision and temperature stability compared to more modern resistor types. Despite their declining use in contemporary electronics due to advancements in resistor technology, carbon composition resistors still find application in specific areas where their unique characteristics are advantageous, making them an essential component in vintage and niche electronic projects.

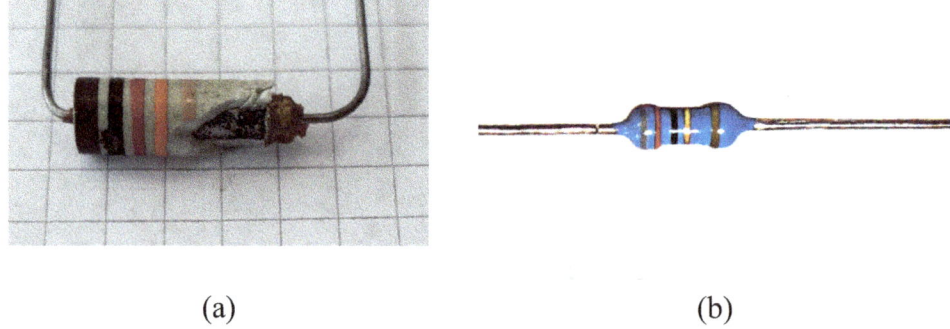

(a)   (b)

Figure 2-1 Various types of through-hole resistors used in electronics along with their characteristics, (a) Carbon composition resistors [3], and (b) Metal Film Resistor

- **Metal Film Resistors:**

Metal film resistors are a common type of fixed resistor used in electronic circuits, see Figure 2-1. They are known for their precision, stability, and low noise characteristics. These resistors are constructed by depositing a thin layer of metal, often nickel-chromium or similar alloys, onto a ceramic substrate. This thin metal layer serves as the resistive element. Metal film resistors come in various resistance values, from fractions of an ohm to several megaohms, and they typically have tight tolerances, such as ±0.1% or ±1%, ensuring that their resistance values are very close to the specified values. These resistors are commonly used in applications where accurate and stable resistance values are crucial, such as in precision amplifiers, analog signal processing, audio equipment, and measurement instruments.

Variable Resistors (Potentiometers):

Potentiometer [4][5]

Figure 2-2 Different types of potentiometer

Their low noise and low temperature coefficient make them ideal for applications requiring minimal interference and high precision. Metal film resistors have become a fundamental component in modern electronics, contributing to the reliability and performance of a wide range of electronic devices and circuits. Variable resistors, commonly known as potentiometers or "pots," are versatile electronic components that allow users to adjust resistance within a specific range. The primary function of a potentiometer is to offer variable resistance. Potentiometers consist of a resistive element, a wiper (contact), and a mechanical shaft or knob (see Figure 2-2). The resistive element is a long, coiled or flat strip of resistive material (typically carbon or conductive plastic) that is mounted inside the potentiometer housing. By turning a knob or shaft, users can change the resistance within a predefined range. The wiper is a movable contact that slides along the resistive element as the shaft or knob is turned. The wiper's position determines the effective resistance between the wiper terminal and the other two terminals. Terminals are provided for connecting the potentiometer in a circuit. Most potentiometers have three terminals: one for the input voltage (fixed end), one for the variable output (wiper), and one for the reference (the other end of the resistive element).

Potentiometers are available in various resistance values and can cover a wide range, from tens of ohms to several megaohms. Typical potentiometer tolerances range from ±5% to ±20%, depending on the precision required for the application.

b. Resistor color code

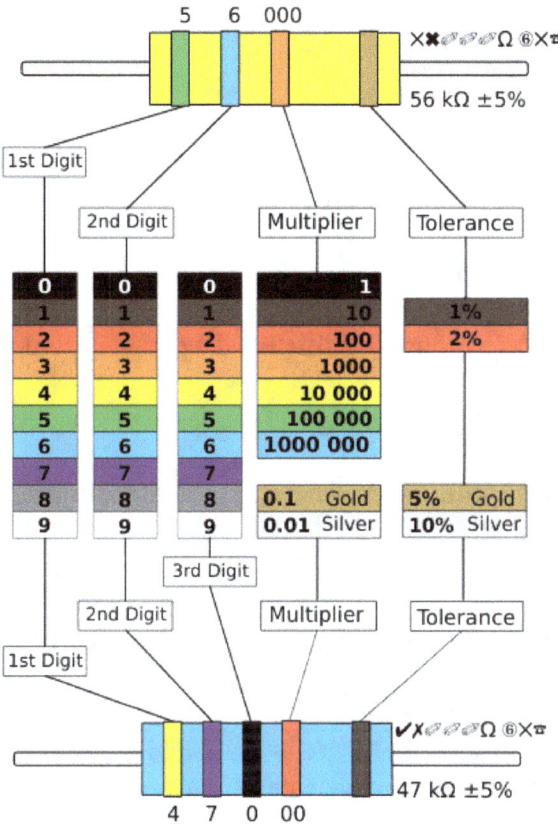

Figure 2-3 Resistor color code for knowing the values of resistors

The resistor color code (Figure 2-3) is used to visually represent the resistance value, tolerance, and sometimes the temperature coefficient of a resistor. It consists of color bands or rings that are typically found on resistors, but it's not commonly used for surface mount resistors (SMD), which often have numerical codes.

Most resistors have either four or five colored bands. Each color corresponds to a digit or a multiplier

1. **First Band (Significant Digit):** The color of the first band represents the first digit of the resistance value.

2. **Second Band (Second Significant Digit):** The color of the second band represents the second digit of the resistance value.

3. **Third Band (Multiplier):** The color of the third band indicates the multiplier by which to multiply the significant digits to get the resistance value. For example, if the third band is red (2), you multiply the significant digits by 10^2 (100).

4. **Fourth Band (Tolerance):** If there is a fourth band, it represents the tolerance of the resistor. The color codes for tolerance are:

   - **Gold (±5%)**

   - **Silver (±10%)**

   - **No Color (±20%)**

   - **Brown (±1%)** (less common)

5. **Fifth Band (Temperature Coefficient, Optional):** Some precision resistors may have a fifth band that represents the temperature coefficient. Common color codes for temperature coefficient are:

- **Brown (100 ppm/°C)**
- **Red (50 ppm/°C)**
- **Orange (15 ppm/°C)**
- **Yellow (25 ppm/°C)**
- **Blue (10 ppm/°C)**

Reading the Value: To read the resistance value, look at the first two bands for the significant digits, the third band for the multiplier, and the fourth band (if present) for the tolerance.

For example, if you see a resistor with bands: Yellow, Violet, Red, Gold, it can be decoded as follows:

- Significant digits: Yellow (4) and Violet (7)
- Multiplier: Red (2)
- Tolerance: Gold (±5%)

So, the resistance value is 47×10^2 ohms, or 4.7 kΩ with a tolerance of ±5%. Now, verify the values of the resistors given in Figure 2-3.

c. What is Surface Mount Device (SMD) resistor?

Resistors (SMD) on Printed   100 ohm SMD [6]   100 ohm SMD

circuit board   [7]

Figure 2-4 SMD resistors soldered on board and parts available in the market

Surface Mount Device (SMD) resistors are compact and versatile electronic components designed for modern electronics manufacturing and miniaturized circuits (Figure 2-4). These resistors are characterized by their small, surface-mountable package that allows them to be soldered directly onto printed circuit boards (PCBs). SMD resistors come in various shapes and sizes, making them suitable for a wide range of applications, from consumer electronics to industrial equipment. Their advantages include space-saving design, excellent thermal performance, and compatibility with automated assembly processes. SMD resistors play a crucial role in contemporary electronics, contributing to the miniaturization and efficiency of electronic devices.

SMD resistors typically have numerical or alphanumeric codes printed on their surface (Figure 2-4).

How to read the SMD Code: You should find a numerical or alphanumeric code printed on SMD resistor surface. This code can vary in format and length depending on the manufacturer and series. Different manufacturers may use slightly different code formats, but generally, the code represents the resistance value in ohms and possibly other information like tolerance and package size.

To determine the specific resistance value of the SMD resistor, you'll need to refer to the datasheet or documentation provided by the manufacturer. The datasheet will include a key or table that explains how to decode the SMD code for that particular series of resistors.

There are also online calculators and databases available that can help you decode SMD resistor codes [8] [9]. You can enter the code, and these tools will provide you with the resistance value and other relevant information. See the links given below:

https://www.utmel.com/tools/smd-resistor-code-calculator?id=33

https://www.digikey.in/en/resources/conversion-calculators/conversion-calculator-smd-resistor-code

II. Capacitors

In section **Error! Reference source not found.**, we learned what capacitor is. When a voltage is applied across the two plates of a capacitor, it stores electrical charge on its plates. When the voltage is removed or reduced, the capacitor discharges and releases the stored energy.

a. What is a Capacitor?

The basic structure of a capacitor consists of two conductive plates separated by a dielectric material (Figure 2-5). The conductive plates are typically made of metal, such as aluminum, tantalum, or copper. The dielectric material is a non-conductive substance placed between the two conductive plates. It is usually made of materials like ceramics, plastics, paper, or electrolytic materials. The dielectric material serves as an insulator, preventing direct electrical contact between the plates. This insulation is essential because it allows the capacitor to store electrical charge without short-circuiting.

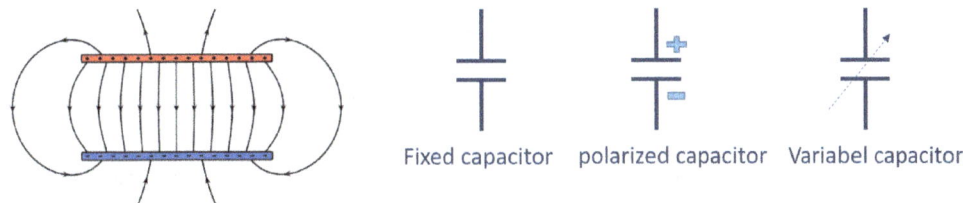

Parallel plate capacitor [10]   Capacitor symbols

Figure 2-5 Structure of a parallel plate capacitor, showing tow plates separated by air as a dielectric

Capacitance is defined as the measure of a capacitor's ability to store electrical charge. It quantifies the amount of electric charge a capacitor can hold for a given voltage across its terminals. The formula for capacitance (C) of a capacitor is:

C = Q / V

Where:

- **C** is the capacitance in farads (F).

- **Q** is the electric charge stored on the capacitor's plates in coulombs (C).

- **V** is the voltage across the capacitor's plates in volts (V).

This formula illustrates that the capacitance of a capacitor is directly proportional to the amount of electric charge it can store for a given voltage. In other words, the higher the capacitance, the more charge the capacitor can hold for a given voltage, making it a more effective energy storage device. The unit of capacitance is the farad (F). One farad is defined as the capacitance of a capacitor that stores one coulomb (C) of electric charge when a voltage of one volt (V) is applied across its terminals:

1 farad (F) = 1 coulomb (C) / 1 volt (V).

Capacitance values are specified on capacitors using their appropriate unit (μF, nF, pF, or F). Figure 2-5 also shows symbols used for fixed, polarized and variable capacitors, discussed below.

b. Types of capacitors

Figure 2-6 shows various types of capacitors used in electronics applications. Each are described below.

1. Electrolytic Capacitors:

Electrolytic capacitors are polarized, meaning they have a positive and negative terminal. They typically have high capacitance values and are available in

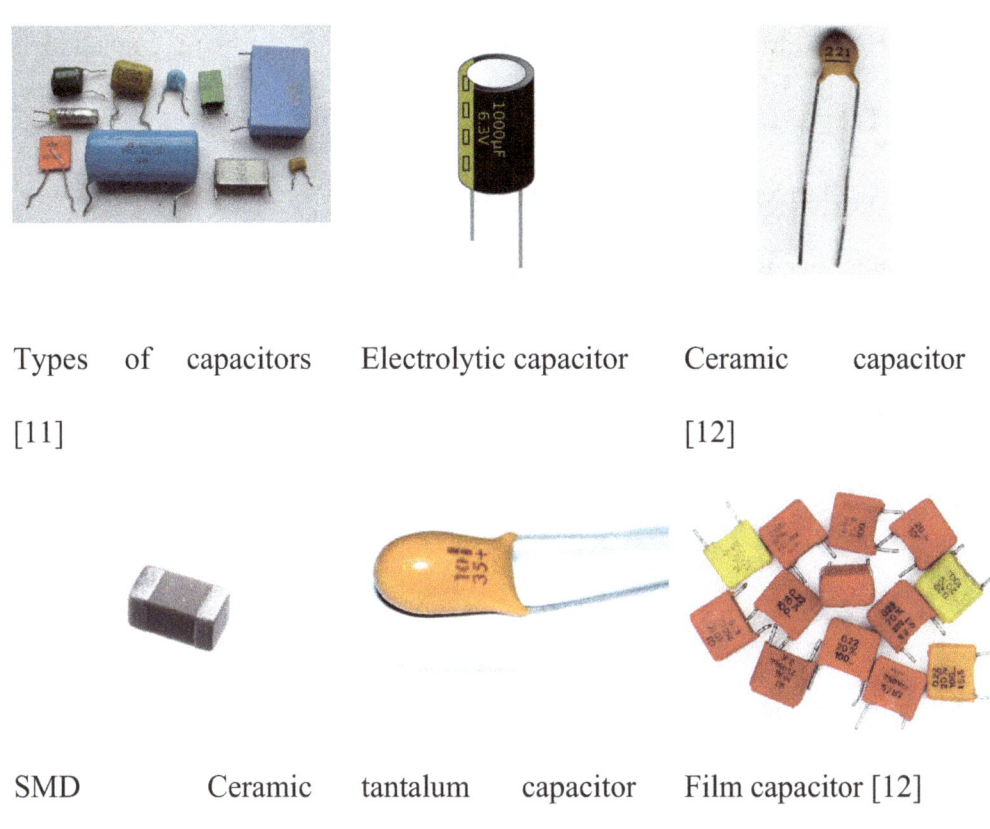

| Types of capacitors [11] | Electrolytic capacitor | Ceramic capacitor [12] |

| SMD Ceramic capacitor [13] | tantalum capacitor [14] | Film capacitor [12] |

Figure 2-6 Types of capacitor used in electronics

cylindrical or rectangular shapes. Aluminum and tantalum are common materials for the electrodes. Electrolytic capacitors are used for high-capacitance, low-frequency applications, such as power supply filtering and audio signal coupling. They are also used in energy storage for applications like camera flashes.

2. Ceramic Capacitors:

Ceramic capacitors are small and come in a wide range of capacitance values. They are made from ceramic materials with metallic electrodes. They are non-polarized. They are suitable for high-frequency applications, decoupling, bypassing, and noise suppression in digital and RF circuits.

3. Tantalum Capacitors:

Tantalum capacitors are polarized and offer a high level of capacitance in a small package. They are known for their stability and reliability. Tantalum capacitors are used in power supply filtering, coupling, and low-frequency timing circuits. They are common in portable electronics and telecommunications equipment.

4. Film Capacitors:

Film capacitors are non-polarized and come in various dielectric materials, including polyester, polypropylene, and polyethylene. They are available in a range of capacitance values. Film capacitors are used in audio applications for signal coupling and filtering. They are also suitable for high-voltage and high-frequency applications.

5. Variable Capacitors:

Variable capacitors have a variable capacitance that can be adjusted manually or electronically. They consist of plates that overlap, and the capacitance changes as the plates move. Variable capacitors are used in tuning circuits for radios, televisions, and antennas. They allow precise tuning of resonant circuits.

6. Mica Capacitors:

Mica capacitors use mica as the dielectric material and have excellent stability and low loss. Mica capacitors are used in high-frequency applications, such as radio transmitters and receivers, and in precision timing circuits.

7. Glass Capacitors:

Glass capacitors use glass as the dielectric material and are known for their stability and low temperature coefficient. Glass capacitors are used in precision timing, tuning, and oscillator circuits.

8. Variable Capacitors:

- The capacitance value of variable capacitors is often indicated on the knob or adjustment mechanism rather than on the capacitor body. It may be represented in picofarads (pF) or another appropriate unit.

c. Reading a capacitor markings and codes

Refer to Figure 2-6.

1. Electrolytic Capacitors:

- Electrolytic capacitors often have a numeric value followed by a unit, such as μF (microfarads) or mF (millifarads). For example, "47μF" means 47 microfarads. Larger capacitors may use codes to represent capacitance values, where a numeric code corresponds to a specific capacitance value.

2. Ceramic Capacitors:

- Ceramic capacitors often use a three-digit code to represent their capacitance value. The first two digits indicate significant figures, and the third digit indicates the number of zeros to add. For example, "104" represents 100,000 pF or 0.1 μF. Figure 2-6 shows a 220pF ceramic capacitor.

3. Tantalum Capacitors:

- Tantalum capacitors typically have a numeric value followed by a unit, such as μF or mF. The value is straightforward to read, as it directly represents the capacitance in microfarads or millifarads.

4. Film Capacitors:

- Film capacitors usually have their capacitance value explicitly marked as a numeric value followed by μF, nF (nanofarads), or pF (picofarads).

The voltage rating of a capacitor is usually indicated as a voltage value (in volts) preceded by a "V" or "WV" (working voltage). For example, "25V" means a voltage rating of 25 volts.

d. How does capacitor charge and discharge?

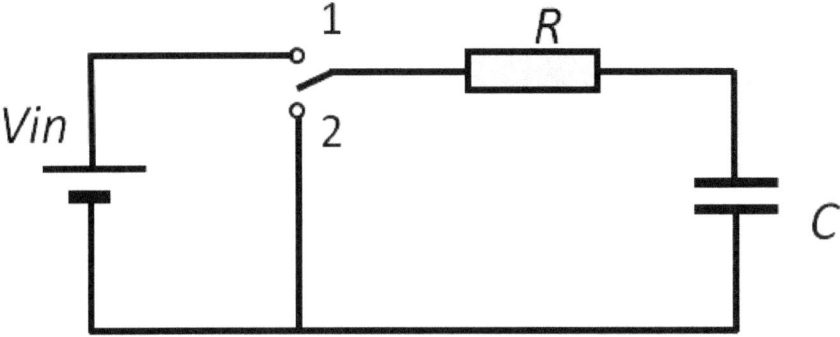

Figure 2-7 Charging and discharging process of a capacitor in a simple RC circuit [12]

Capacitors charge and discharge in response to changes in voltage according to certain principles. Let's explore how capacitors behave during charging and discharging: (Figure 2-7)

1. Charging of Capacitors:

- When a voltage source (e.g., a battery) is connected across the terminals of a capacitor, the capacitor begins to charge.

- Electrons from the negative terminal of the voltage source flow onto the capacitor's plate connected to the negative terminal, while electrons from the other plate flow away to the positive terminal of the source.

- As the process continues, the voltage across the capacitor increases, and the rate of charge accumulation decreases over time.

- In an ideal scenario with no resistance in the circuit, a capacitor will charge to the source voltage in a theoretically infinite amount of time.

2. Discharging of Capacitors:

- When a charged capacitor is disconnected from its voltage source and connected to a circuit with a lower voltage or a resistor, it begins to discharge.

- Electrons flow from one plate of the capacitor through the circuit (or resistor) to the other plate, gradually reducing the voltage across the capacitor.

- Similar to charging, the rate of discharge decreases over time.

- In an ideal scenario with no resistance in the circuit, a capacitor will take an infinite amount of time to fully discharge.

The RC Time Constant (τ):

In practical circuits, there is often some resistance in the circuit due to the wires, connections, and sometimes intentional resistors (Figure 2-7). This resistance affects the rate at which a capacitor charges and discharges. The time it takes for a capacitor to charge to approximately 63.2% of its final voltage (or discharge to 36.8% of its initial voltage) in the presence of resistance is defined by the RC time constant (τ). The RC time constant is calculated using the following formula:

$$\tau = R \times C$$

Where:

- τ **(tau)** is the time constant in seconds (s).

- **R** is the resistance in ohms (Ω) in the circuit.

- **C** is the capacitance in farads (F) of the capacitor.

The RC time constant characterizes the time it takes for a capacitor to reach about 63.2% of its final voltage during charging or discharge. It also indicates how quickly a capacitor responds to changes in voltage. A smaller τ results in a faster response, while a larger τ leads to a slower response.

III. Inductors

let's understand what an inductor is. An inductor is a two-terminal passive electrical component that stores energy in the form of a magnetic field (Figure 2-8). When an electric current passes through the inductor, it produces a surrounding magnetic field. If the current varies with respect to time, it creates a time-varying magnetic field around it. This phenomenon is known as electromagnetic induction, based on Faraday's law. As a result of this action, an electromotive force (EMF), or voltage, is induced across the two terminals of the inductor. The polarity of this EMF opposes the direction of the current that produces it. Mathematically, the voltage across the inductor is equal to the inductance (L) multiplied by the rate of change of the current flowing through it.

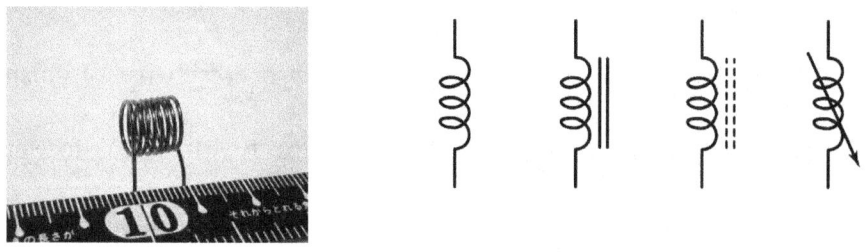

Inductor [15]   Symbols [16]

Figure 2-8 Inductor and its symbols

In other words, the value of the inductance, denoted as L, can be defined as the ratio of the voltage across it divided by the rate of change of current, which is expressed as dI/dt. Typically, inductors are made by winding insulated wire into a coil, and there are various types of inductors available, with values ranging from a few mH to as high as 20 Henry, or even in the mH range for smaller values. Mathematically, an inductance (L) is defined as the ratio of magnetic flux (Φ_B) to current (I), expressed as $L = \Phi_B / I$. When an electromotive force (EMF) is generated across the inductor, represented as V_L, it is precisely the rate of change of magnetic flux with a negative sign, $V_L = -d\Phi_B/dt$. Importantly, magnetic flux (Φ_B) is fundamentally a product of inductance (L) and current (I), $\Phi_B = L * I$. By substituting this value into the expression for V_L, we arrive at the final mathematical representation: VL = - L * (dI/dt). These equations offer a mathematical framework for understanding how inductors store energy in magnetic fields and respond to changes in current, ultimately producing induced EMFs in electrical circuits.

Chapter 3   Electronic Fundamentals

I.   Alternating Current (AC) and Direct Current (DC) in Electricity

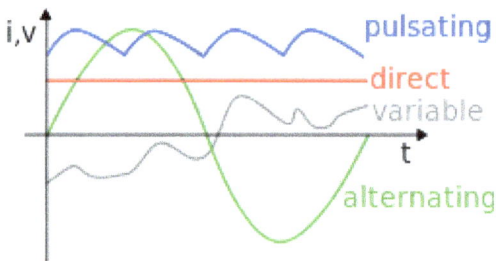

Figure 3-1 AC and DC, The horizontal axis measures time (it also represents zero voltage/current); the vertical, current or voltage [17].

Alternating Current (AC) and Direct Current (DC) are two fundamental types of electric currents with distinct properties and applications. AC is characterized by its continuous change in direction, regularly shifting back and forth. It is the form of electricity commonly used for delivering power to homes and businesses. When you plug in your kitchen appliances, televisions, fans, or electric lamps, they typically run on AC. The AC waveform is often represented as a sine wave, which is a smooth, wavy line. In contrast, DC flows steadily in one direction, as seen in batteries or the power source for many electronic devices. While AC is versatile and efficient for long-distance power transmission, DC is more stable and reliable for electronic circuits. Both types have their unique strengths and are used in various applications, making them essential components of our modern electrical systems.

II. Voltage, current, and resistance

Voltage, current, and resistance are the fundamental building blocks of electronics, and understanding these concepts is key to understand electronic circuits.

Voltage (V):

Voltage, often denoted as "V," is a fundamental electrical quantity in electronics. It represents the electric potential difference between two points in an electrical circuit. In simpler terms, voltage can be thought of as the "electrical pressure" that pushes electric charge (usually in the form of electrons) through a conductor like a wire or an electronic component. Figure 3-2 shows symbol of a voltage source, various sources and checkpoints where voltage is available. It also shows an application circuit where voltage can be applied and measured. It also shows electronic power supply that provides voltage to other circuits.

High voltage electric station

High voltage symbol

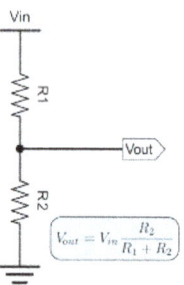

Voltage divider in electrical circuit

Battery

Voltage source unit

Power supply unit

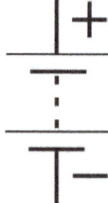

Symbol of a voltage

Figure 3-2 Sources of voltage, symbol, circuit application and electronic

unit

Facts about voltage:

- In an electrical circuit, voltage provides the force that moves electric charge (current) from one point to another.

- Voltage can be supplied by various sources, including batteries, power supplies, generators, and electrical outlets. These sources create a potential difference, causing electric charge to flow when a circuit is completed.

- Voltage is always measured relative to a reference point. When we say an object has a voltage of, for example, 5 volts, it means it has 5 volts more (or less) electrical potential energy than the chosen reference point.

- Voltage levels can signal "on" or "off" states in digital circuits and control the brightness of a light-emitting diode (LED).

- Kirchhoff's Voltage Law (KVL) (discussed later) is a fundamental principle in circuit analysis. It states that the sum of the voltages around any closed loop in a circuit is equal to zero, emphasizing the conservation of energy in electrical circuits.

- Voltage can pose electrical hazards, so it's essential to handle electrical equipment and circuits with care. High-voltage circuits can cause electrical shocks or damage to components.

Current (I):

Current, denoted as "I," is a fundamental electrical quantity in electronics and represents the flow of electric charge through a conductor, such as a wire or a circuit. It is the rate at which electric charge moves past a given point in a circuit and is measured in amperes (A). See Figure 3-3 for facts explained below.

Facts about current:

- Current is the flow of electric charge, typically in the form of electrons, through a conductive pathway.

- Current is measured in amperes (A). One ampere is defined as one coulomb of charge passing through a conductor per second.

- The direction of current flow is conventionally defined as the direction in which positive charges would move. In reality, electrons, which are negatively charged, flow in the opposite direction.

- Direct Current (DC): In DC circuits, current flows steadily in one direction.

- Alternating Current (AC): In AC circuits, current reverses direction periodically, typically in a sinusoidal waveform. AC is commonly used in household electricity.

- Ohm's Law (explained below) is a fundamental principle in electronics, it relates voltage (V), current (I), and resistance (R) in a circuit. It is expressed as V = I x R, where voltage is directly proportional to current and resistance.

- The principle of current conservation states that in a closed circuit, the total current entering a point is equal to the total current leaving that point. This principle ensures that charge is conserved within the circuit.

- High current levels can pose electrical hazards, such as electrical shocks and circuit overheating. Proper safety precautions are necessary when working with electrical currents.

- Current can be provided by various sources, including batteries, generators, and power supplies. These sources create a potential difference (voltage) that drives current through a circuit.

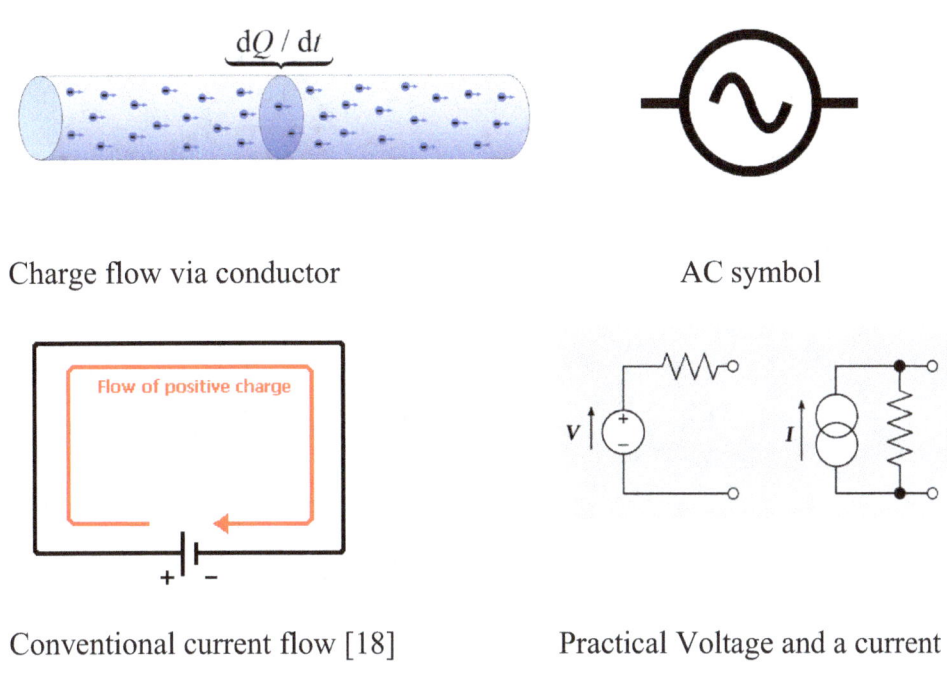

Charge flow via conductor   AC symbol

Conventional current flow [18]   Practical Voltage and a current

sources [19]

Figure 3-3 symbol for a current, flow of charges, symbol for a AC source,
direction of current and sources of current (that produces a voltage) and
voltage (that produces a current when load is connected)

Resistance (R):

Resistance, often denoted as "R," is a fundamental electrical property that
describes the opposition to the flow of electric current in a conductor, such
as

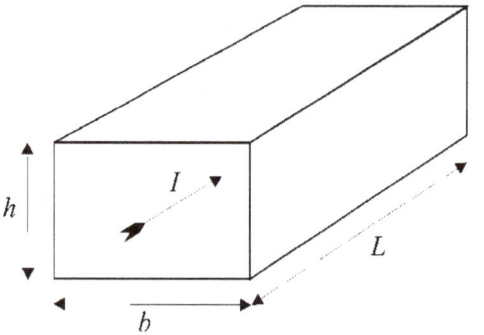 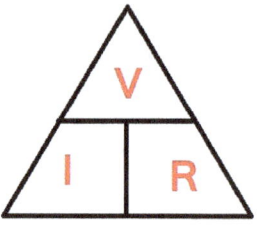

Resistance of a material [20]　　　　　Ohm's law depicting relation

between voltage, current and

resistance [21]

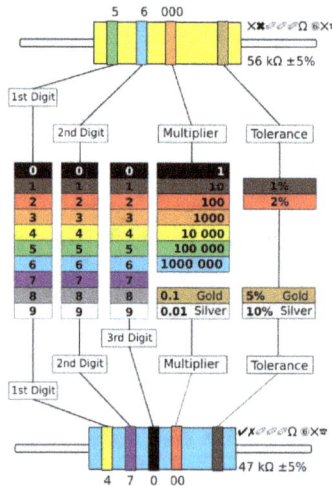

Resistor colour code

Figure 3-4 Resistance of a material, ohm's law showing relation between

current, voltage and resistance and color code of a resistor

a wire, component, or circuit. It is a crucial concept in electronics and is measured in ohms (Ω).

Facts about resistance:

- Resistance represents the degree to which a material or component resists the flow of electric current.

- Resistance is measured in ohms (Ω). One ohm is defined as the resistance that allows a current of one ampere (A) to flow when a voltage of one volt (V) is applied across it.

- Different materials have different inherent resistances. For example, metals like copper have low resistance and are good conductors, while materials like rubber have high resistance and are insulators.

- Longer conductors generally have higher resistance. A wider conductor has lower resistance.

- The resistance of some materials, especially semiconductors, can change with temperature.

  Ohm's Law relates voltage (V), current (I), and resistance (R) in a circuit. It is expressed as $V = I \times R$, indicating that voltage is directly proportional to current and resistance.

- When current flows through a resistor, it generates heat due to the resistance. The power dissipated as heat in a resistor is calculated using the formula $P = V^2 / R$, where P is power, V is voltage, and R is resistance.

To identify the resistance value of a resistor, a color code system is often used, where different colored bands on the resistor correspond to specific resistance values.

- Resistance is a fundamental property in various electronic applications, such as in voltage dividers, filters, and current-limiting circuits.

Some components, like potentiometers and variable resistors, allow the resistance value to be adjusted, making them useful for applications where resistance needs to be variable. See Figure 3-4.

III.   Ohm's Law and how to apply it

Ohm's Law is a fundamental principle in electronics, named after the German physicist Georg Simon Ohm. It describes the relationship between voltage (V), current (I), and resistance (R) in an electrical circuit. See Figure 3-4. Ohm's Law is expressed mathematically as $V = I \times R$, where:

V represents voltage in volts (V).

I represents current in amperes (A).

R represents resistance in ohms (Ω).

Ohm's Law tells us that the voltage across a component in a circuit is directly proportional to the current flowing through it. If you increase the voltage, the current increases proportionally, assuming the resistance remains constant.

Ohm's Law allows you to calculate any one of the three variables (voltage, current, or resistance) if you know the values of the other two. This is useful in designing and analyzing circuits. For example, if you have a resistor with a known resistance (R) and a known current (I) passing through it, you can find the voltage drop (V) across the resistor using V = I x R.

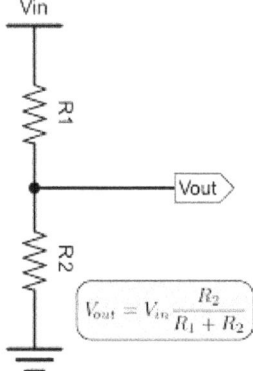

Figure 3-5 Voltage divider circuit

Ohm's Law is fundamental in designing voltage divider circuits (see voltage divider circuit in Figure 3-5. By connecting resistors in series, you can create a voltage divider that provides a specific voltage output relative to the input voltage. This is commonly used in applications like adjusting brightness in LED displays or setting reference voltages. Ohm's Law is essential for current limiting in circuits to protect components or prevent excessive current flow. By choosing the appropriate resistance value, you can limit the current passing through a component to a safe level. Engineers and hobbyists use Ohm's Law extensively when analyzing circuits. It helps determine how voltages and currents behave in different parts of a circuit, making it a valuable tool for troubleshooting and optimizing circuit performance.

Ohm's Law also plays a role in power calculations. You can use it to find the power dissipated in a resistor using the formula $P = V \times I$ or $P = I^2 \times R$. These equations are crucial for selecting components that can handle the power generated in a circuit.

Ohm's Law is foundational in a wide range of electronic applications, including power supplies, amplifiers, lighting circuits, and more. It ensures that circuits operate as intended and within safe limits.

IV. Electrical power and energy

Understanding electrical power and energy is essential in electronics and electrical engineering. These concepts are fundamental for designing circuits, calculating energy consumption, and optimizing the efficiency of electrical systems. (see High voltage electric station in **Error! Reference source not found.**)

Electrical Power (P):

Electrical power, denoted as "P," is the rate at which electrical energy is consumed, generated, or transferred in an electrical circuit. It quantifies how quickly electrical work is done.

The formula for electrical power is given by $P = V \times I$, where:

P represents power in watts (W).

V represents voltage in volts (V).

I represents current in amperes (A).

Facts:

- Power is measured in watts (W), and it indicates the rate of energy transfer or consumption per unit of time.

- In a circuit, power represents the ability to do work, such as lighting a bulb or running a motor.

- High-power devices consume more electrical energy per unit of time than low-power devices.

- Power is a crucial factor in designing circuits, selecting components, and calculating power losses.

- Power ratings are often used to specify the capacity or performance of electrical devices. For example, a 100W light bulb consumes electrical power at a rate of 100 watts when lit.

- Engineers use power to assess the efficiency of electrical systems. By comparing input and output power, they can determine how effectively energy is converted and used.

Electrical Energy (E):

Electrical energy, denoted as "E," is the total amount of work done or energy consumed by an electrical device or system over a period of time. It is measured in watt-hours (Wh) or kilowatt-hours (kWh).

Electrical energy can be calculated using the formula $E = P \times t$, where:

E represents energy in watt-hours (Wh) or kilowatt-hours (kWh).

P represents power in watts (W).

t represents time in hours (h).

Facts:

- Electrical energy is a cumulative measure of power consumption over time. It is what you are billed for on your electricity bill.

- 1 kilowatt-hour (kWh) is equivalent to 1,000 watt-hours (Wh). It's a common unit for measuring electrical energy in residential and commercial settings.

- Electrical energy is used to assess the long-term cost of operating electrical devices and systems.

- Assessing energy consumption helps evaluate the environmental impact of electrical systems, as it relates to carbon emissions and sustainability efforts.

- Engineers use energy calculations to design efficient systems and optimize energy usage in various applications, from renewable energy systems to electric vehicles.

V.   Kirchhoff's laws as circuit analysis techniques

In this module, we will explore a fundamental law in network analysis known as Kirchhoff's law, named after the German scientist Mr. Kirchhoff.

There are two Kirchhoff's laws: Kirchhoff's current law (KCL) and Kirchhoff's voltage law (KVL). First, we will focus on Kirchhoff's current law, often abbreviated as KCL.

a. Kirchhoff's current law (KCL)

Consider a network or circuit, as shown (Figure 3-6), with currents labeled as I_1, I_2, I_3, and I_4, each with its specified direction. The central point, represented as a black dot, is known as a node or junction within the network.

Kirchhoff's current law (KCL) states that the algebraic sum of the currents entering a specific node is equal to the sum of the currents exiting the same node. In other words, the sum of currents flowing into a node equals the sum of currents flowing out of that node.

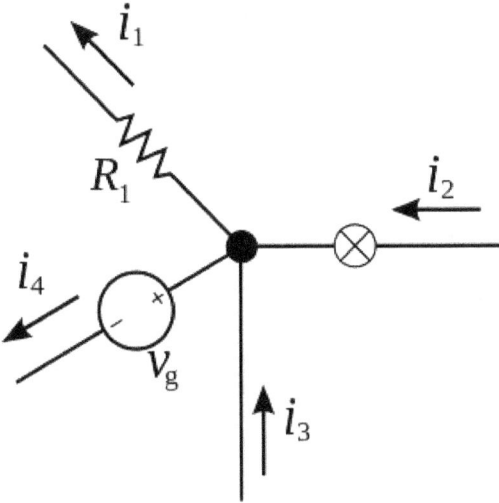

Figure 3-6 Kirchhoff's current law [22]

For instance, at this node, we observe that I_2 and I_3 are entering. Therefore, we can express KCL as:

$I_2 + I_3 = I_1 + I_4$.

This equation signifies that the sum of currents entering (I_2 and I_3) equals the sum of currents exiting (I_1 and I_4) the node.

Alternatively, we can express KCL as the algebraic sum of all currents in a network being equal to zero:

$I_2 + I_3 - I_1 - I_4 = 0$.

This equation reinforces that in a network or circuit with conductors, the algebraic sum of all currents is always zero, representing the conservation of electric charge.

Kirchhoff's current law (KCL) is applicable to both linear and non-linear circuits, making it a fundamental principle in network analysis.

b. Kirchhoff's voltage law (KVL)

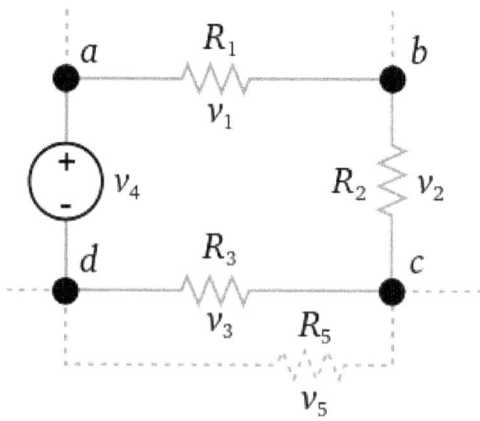

Figure 3-7 Kirchhoff's voltage law [22]

In this module, we are going to understand the second law from Kirchhoff, called Kirchhoff's voltage law, or KVL for short. Imagine you have a network or a circuit as shown in the figure. It consists of a voltage source known as V_4, resistors R_1, R_2, R_3, and another resistor R_5. Additionally, you have black dots labeled as a, b, c, and d, which are referred to as nodes or junctions within the circuit.

In the network shown in Figure 3-7, we have voltage drops across R_1 (indicated as V_1), R_2 (indicated as V_2), R_3 (indicated as V_3), and R_5 (indicated as V_5) due to the current flowing through these resistors as a result of the voltage source. According to Ohm's law, these voltage drops occur.

Since R_3 and R_5 are connected in parallel, as both of their ends are connected to the same nodes or junctions, the voltage drop across each of them is the same. In other words, V_3 and V_5 are equal.

Our focus in this module is to explain Kirchhoff's voltage law, or KVL. This law states that the algebraic sum of all the voltages around a closed loop is equal to 0. Mathematically, we can express KVL as:

$V_1 + V_2 + V_3 + V_4 = 0.$

This equation defines KVL, where V_3 and V_5 are considered equal.

KVL can be applied to any circuit, whether it is linear or non-linear, consists of only passive elements, active elements, or a combination of both.

c. Example 1

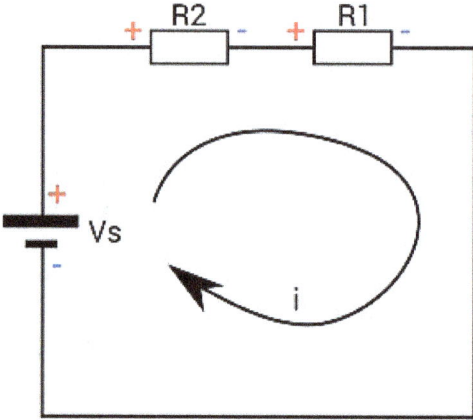

Figure 3-8 Example 1 circuit [23]

In this network (Figure 3-8), we can indeed apply Kirchhoff's Voltage Law (KVL). KVL states that the algebraic sum of all the voltages around a closed loop is equal to zero. Allow me to explain how this applies to the given linear circuit, which consists of a voltage source and two resistors in series.

Firstly, we have the voltage source, denoted as V_S. Due to this voltage source, a current, denoted as I, flows through the circuit. This current passes through both R_1 and R_2 and returns to the negative terminal of the source. The voltage drop across R_2 due to the current I is denoted as VR_2, and the voltage drop across R_1 is denoted as VR_1.

Now, let's consider the voltages present in the circuit: V_S, VR_2, and VR_1.

According to KVL, we can express it as:

$V_S + VR_2 + VR_1 = 0$.

This equation represents Kirchhoff's Voltage Law, stating that the sum of all

voltages around a closed loop is equal to zero.

d. **Example 2**

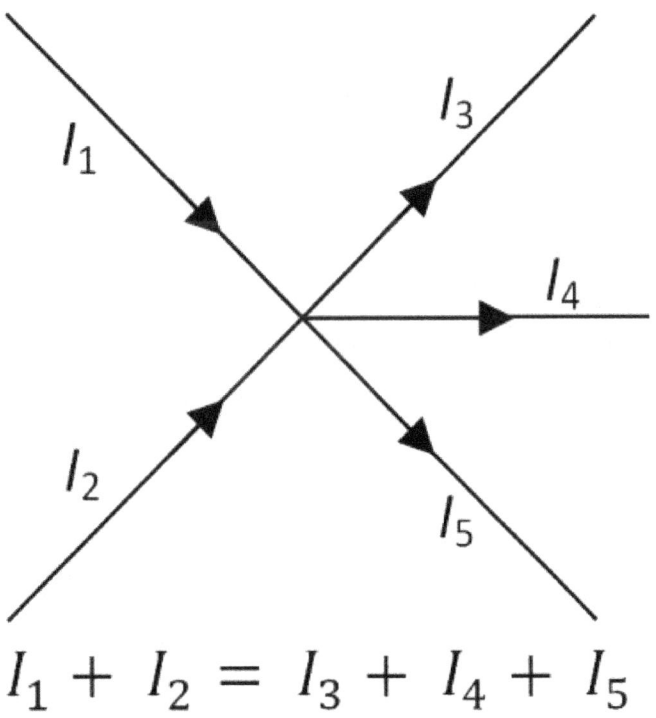

Figure 3-9 Example circuit 2 [24]

In this circuit (Figure 3-9), we are going to apply Kirchhoff's Current Law (KCL). You can see that there are currents denoted as i_1, i_2, i_3, i_4, and i_5. This specific point is referred to as a node or junction. It's important to note that this point is called a node or junction.

KCL states that in a network, as shown in this figure, the sum of the currents entering a node must equal the sum of the currents exiting the same node.

For this node, we have i1 and i2 entering, so we can express this as:

$i_1 + i_2$

On the exiting side, we have i_3, i_4, and i_5, therefore,

i1 + i2 = i3 + i4 + i5.

This equation demonstrates Kirchhoff's Current Law in action.

VI.   Series and parallel circuits

In electronics, there are two ways to connect components: series and parallel. When components are in series (Figure 3-10), they form a single path for electric current, and the current is the same through each component. In parallel, components have multiple paths, and the voltage is the same across each one.

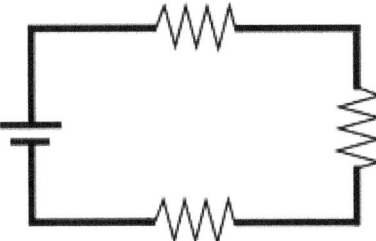

Figure 3-10 A series circuit with a voltage source (such as a battery, or a cell) and three resistance units [25]

Imagine a simple circuit with four light bulbs and a 12-volt battery. If the bulbs are connected one after the other in a loop, they are in series. In this case, each bulb only gets 3 volts, which might not make them glow. However, if the bulbs are connected separately to the battery, they are in parallel. Now, each bulb gets the full 12 volts, and they all glow.

The key difference is that in a series circuit, if one component fails, the whole circuit breaks. But in a parallel circuit, if one component fails, the others can still work. This is why parallel circuits are commonly used in homes and devices to ensure that if one part stops working, the rest can still function.

a. Series circuits

In a series circuit, the **current** is the same for all of the components, and it can be calculated using the following equation:

$$I_{total} = I_1 = I_2 = ... = I_n$$

Where:

I_{total} is the total current in the series circuit.

I_1, I_2, ..., I_n are the currents through each individual component in the series circuit.

In a series circuit, the **total voltage** (V_{total}) across the circuit is equal to the sum of the voltages across each individual component. This can be expressed using the equation:

$$V_{total} = V_1 + V_2 + ... + V_n$$

Where:

V_{total} is the total voltage in the series circuit.

V_1, V_2, ..., V_n are the voltages across each individual component in the series circuit.

In a series circuit, the **total resistance** (R_{total}) is equal to the sum of the resistances of the individual components. You can express this using the equation:

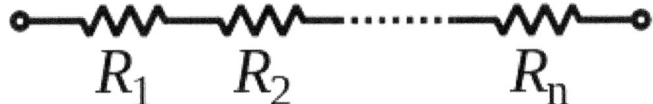

Figure 3-11 Resistance in series [25]

$R_{total} = R_1 + R_2 + ... + R_n$

Where:

R_{total} is the total resistance in the series circuit.

$R_1, R_2, ..., R_n$ are the resistances of each individual component in the series circuit.

Conductance (G) is the reciprocal of resistance in an electrical circuit. In a series circuit, the total conductance (G_{total}) of pure resistances is calculated by the sum of the reciprocals of the individual conductances. The equation for calculating G_{total} in a series circuit is:

$1/G_{total} = 1/G_1 + 1/G_2 + ... + 1/G_n$

Where:

G_{total} is the total conductance in the series circuit.

G_1, G_2, ..., G_n are the conductances of each individual component in the series circuit.

Conductance is typically measured in siemens (S), which is the reciprocal of ohms (Ω). It represents how easily electric current can flow through a component or circuit.

In a series circuit, **inductors** follow the same law as resistors and capacitors.

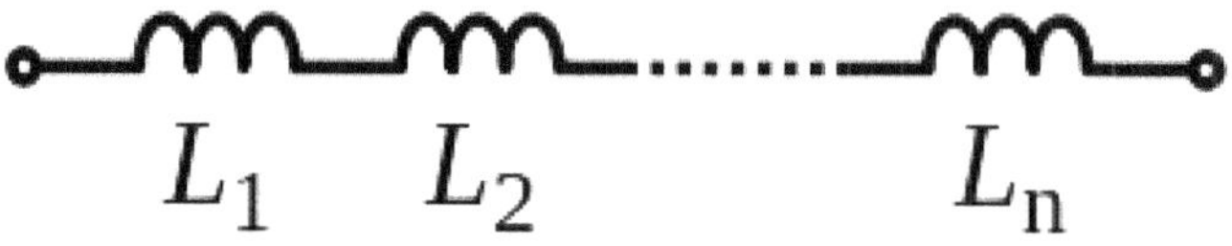

Figure 3-12 Inductance in series [25]

The total inductance (L_{total}) of non-coupled inductors in series is equal to the sum of their individual inductances. The equation for calculating L_{total} in a series circuit is:

$$L_{total} = L_1 + L_2 + ... + L_n$$

Where:

L_{total} is the total inductance in the series circuit.

L_1, L_2, ..., L_n are the inductances of each individual inductor in the series circuit.

Inductance (L) is a property of inductors, and it is measured in henrys (H). It represents the ability of an inductor to store electrical energy in the form of a magnetic field when current flows through it. In a series circuit, the total inductance is the sum of the individual inductances.

In a series circuit, **capacitors** also follow the same law as resistors and inductors.

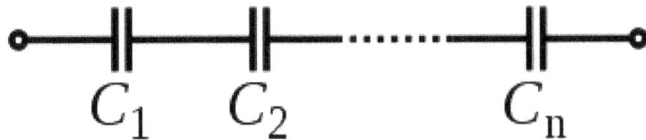

Figure 3-13 Capacitance in series [25]

The total capacitance (C_{total}) of capacitors in series is equal to the reciprocal of the sum of the reciprocals of their individual capacitances. The equation for calculating C_{total} in a series circuit is:

$$1 / C_{total} = 1 / C_1 + 1 / C_2 + ... + 1 / C_n$$

Where:

- C_{total} is the total capacitance in the series circuit.

- $C_1, C_2, ..., C_n$ are the capacitances of each individual capacitor in the series circuit.

Capacitance (C) is a property of capacitors, and it is measured in farads (F). It represents the ability of a capacitor to store electrical energy in the form of an electric field when a voltage is applied across it. In a series circuit, the total capacitance is calculated by adding the reciprocals of the individual capacitances and taking the reciprocal of the sum.

When two or more **switches** are connected in series, they form a logical AND gate in electronics. This means that for current to flow through the circuit, all the switches must be closed (in the ON position). If any of the switches in the series circuit is open (in the OFF position), it will interrupt the flow of current through the entire circuit.

Cells and batteries are often connected in series to achieve higher voltages. A battery is essentially a collection of electrochemical cells. When these cells are connected in series, the total voltage of the battery is equal to the sum of the individual cell voltages.

For example, let's consider a 12-volt car battery. It contains six 2-volt cells connected in series. When these cells are connected this way, their voltages add up to provide a total voltage of 12 volts for the battery. This higher voltage is essential to power various electrical components in a car.

In some cases, such as trucks, two 12-volt batteries may be connected in series to create a 24-volt system. When batteries are connected in series, their voltages are additive, which makes them suitable for applications requiring higher voltage levels.

This series connection of cells or batteries is a common technique to meet specific voltage requirements in various electrical and electronic devices.

b. **Example:**

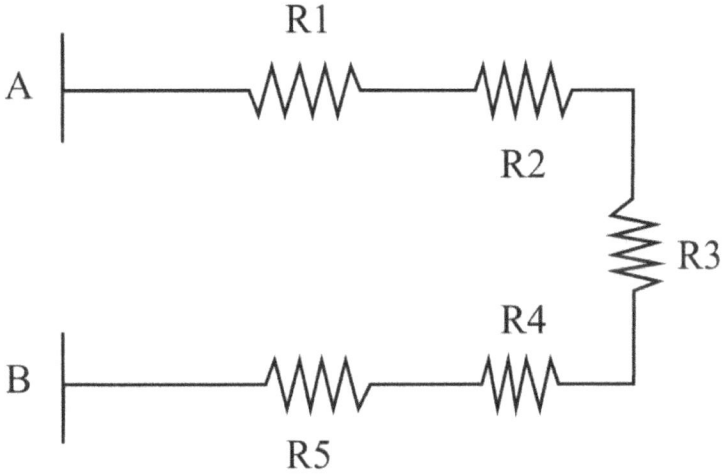

Figure 3-14 Series connection of resistors [26]

Do you know how to determine the effective resistance of network in Figure 3-14. When I refer to effective resistance, I mean the total or equivalent resistance from point A to point B. We're examining this network configuration, which shows that resistors R1, R2, R3, R4, and R5 are connected in series. Therefore, the total resistance from point A to point B can be calculated as the sum of R1, R2, R3, R4, and R5. If each of these resistors has a value of 1 kilo ohm, then the total resistance would be 5 kilo ohms. This clearly demonstrates a series connection of the resistors.

$R_{total} = R1 + R2 + R3 + R4 + R5$

In this case, if each resistor has a value of 1 kilo ohm, the total resistance (R_{total}) would be:

$R_{total} = 1\ k\Omega + 1\ k\Omega + 1\ k\Omega + 1\ k\Omega + 1\ k\Omega = 5\ k\Omega$

c. Parallel circuits

In a parallel circuit, components are connected in multiple paths, and each component has the same voltage across it, equal to the voltage across the network. The current through the network is equal to the sum of the currents through each component.

Voltage (V):

In a parallel circuit, the voltage (V) is the same across all components.

$V_{total} = V1 = V2 = V3 = ... = Vn$

Current (I):

The total current (Itotal) is the sum of the currents through each branch or component in parallel.

$I_{total} = I1 + I2 + I3 + \dots + In$

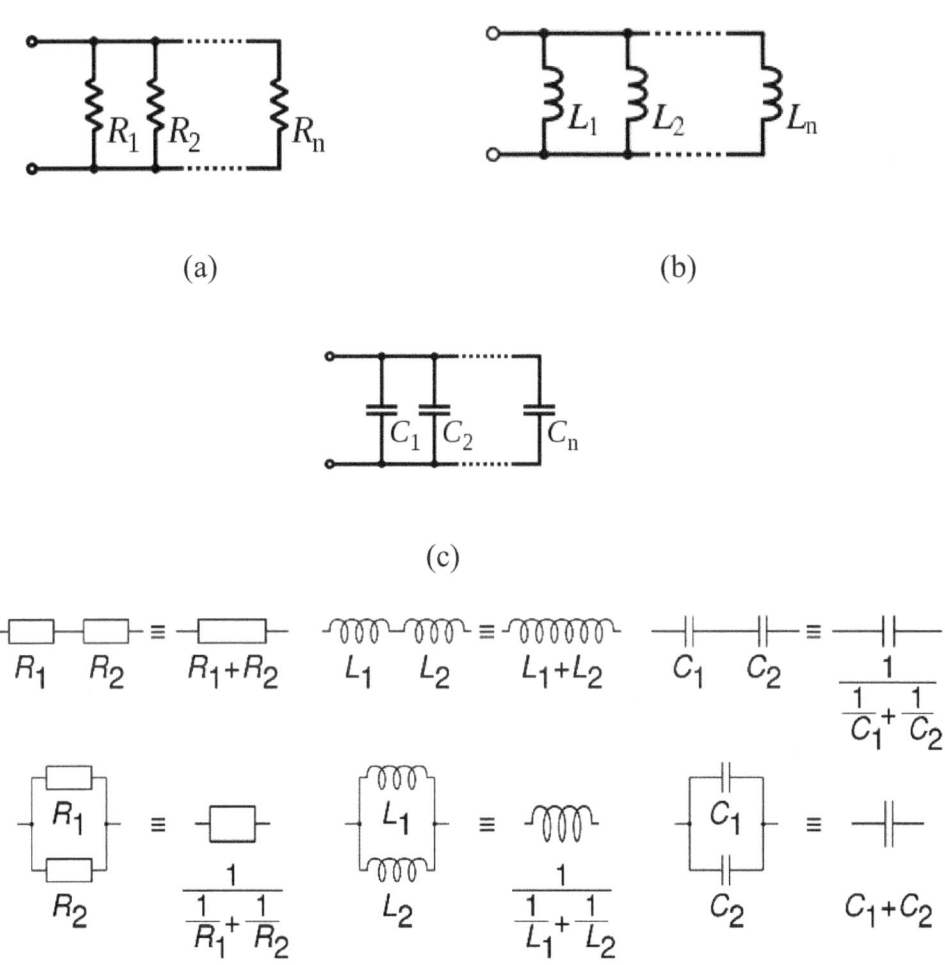

(a)

(b)

(c)

(d)

Figure 3-15 parallel circuits for resistive, capacitor and inductor network, (a to c) and final summary of series and parallel combination of passive components in electrical network [27]

Figure 3-15 shows parallel circuit for resistors, capacitors, inductors, and finally also provides a summary of series and parallel circuits of the passive components.

Resistance (R):

In a parallel circuit, the reciprocal of the equivalent resistance ($1/R_{\_total}$) is the sum of the reciprocals of the individual resistances.

$$1 / R_{\_total} = 1 / R1 + 1 / R2 + 1 / R3 + ... + 1 / Rn$$

Capacitors (C):

In a parallel circuit, the total capacitance (C_total) is the sum of the individual capacitances.

$$C_{\_total} = C1 + C2 + C3 + ... + Cn$$

Inductors (L):

For non-coupled inductors in parallel, the total inductance ($L_{\_total}$) is the sum of the individual inductances.

Equation for Total Inductance ($L_{\_total}$):

$$L_{\_total} = L1 + L2 + L3 + ... + Ln$$

Switches:

In a parallel circuit, two or more switches in parallel form a logical OR. The circuit carries current if at least one switch is closed.

Cells and Batteries:

In a parallel configuration, the voltage of the battery remains the same, and the total current capacity is the sum of the individual cells.

Equation for Total Voltage ($V_{\_total}$):

$$V_{\_total} = V1 = V2 = V3 = ... = Vn$$

d. Example:

Determine the equivalent network of the circuit at its two open ends.

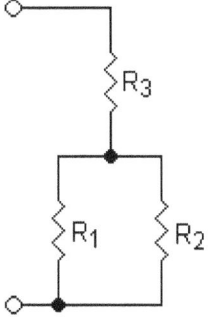

Figure 3-16 Resistive network [27]

83

To determine the equivalent resistance of the network between its two open points, let's break it down:

1. Resistors R1 and R2 are in parallel to each other because both ends of R1 connect to both ends of R2. The equivalent resistance (Rp) of two resistors in parallel can be calculated using the formula:

$1/Rp = 1/R1 + 1/R2$

2. Now, resistor R3 is in series with the parallel combination of R1 and R2. The equivalent resistance (Rs) of resistors in series is simply the sum of their resistances:

$Rs = R3 + Rp$

So, the total equivalent resistance (R_{total}) of the network between the two open points is:

$R_{total} = R3 + (1 / (1/R1 + 1/R2))$

If you have specific resistance values for R1, R2, and R3, you can substitute those values into the equation to calculate the equivalent resistance.

VII. Voltage and current dividers

a. Voltage division

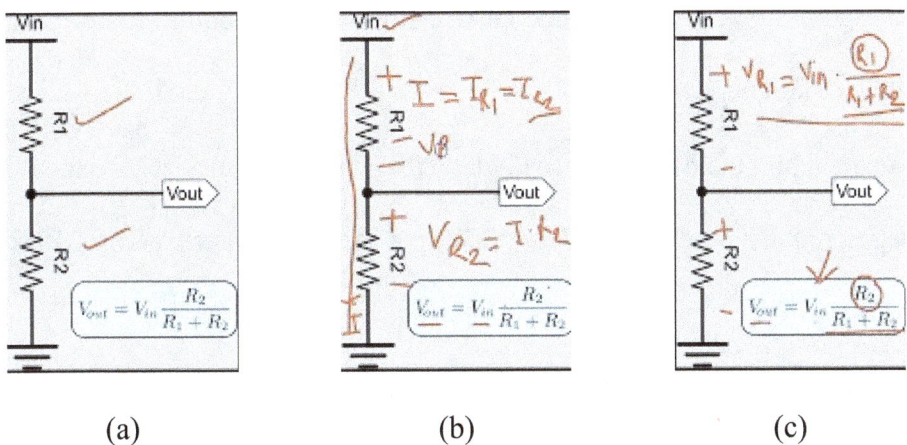

(a)   (b)   (c)

Figure 3-17 voltage division

Let's understand how the voltage division circuit works. Imagine you have

two resistors, R1 and R2, connected in series (Figure 3-17 a). You apply an

input voltage, Vin, to one end of R1. At the junction of R1 and R2, you tap

into the circuit and measure the output voltage, Vout. Mathematically, Vout

can be expressed as:

$$Vout = Vin \times (R2 / (R1 + R2))$$

Essentially, we're measuring the voltage across resistor R2, so Vout is also

the voltage across R2, denoted as VR2.

Now, when you apply a voltage to this circuit, a current, I, flows from the supply towards the ground. This current, I, flows through both R1 and R2 because they are connected in series. You can express the current as $I_{r1} = I_{r2} = I$ (Figure 3-17 b). So, the same current flows through both R1 and R2 due to their series connection.

We can apply Ohm's law to calculate the voltage across R2. Using the equation $V_{out} = V_{in} \times (R2 / (R1 + R2))$, we can also express VR2 as $V_{R2} = I \times R2$.

Similarly, if you want to calculate the voltage across R1, you can use $V_{R1} = I \times R1$, which is a straightforward application of Ohm's law. If you want to calculate the voltage across R1 using the voltage division rule, it can be expressed as $V_{R1} = V_{in} \times (R1 / (R1 + R2))$ (Figure 3-17 c).

Comparing these equations, you'll notice that whether you calculate Vout, VR2, or VR1, you select the resistor across which you want to determine the voltage and place its resistance in the numerator. In the denominator, you use the sum of all the resistors in the circuit.

This concept is known as the voltage division rule, and circuits like this are called voltage dividers. We've calculated voltages and currents using Ohm's law and the voltage division rule.

Example:

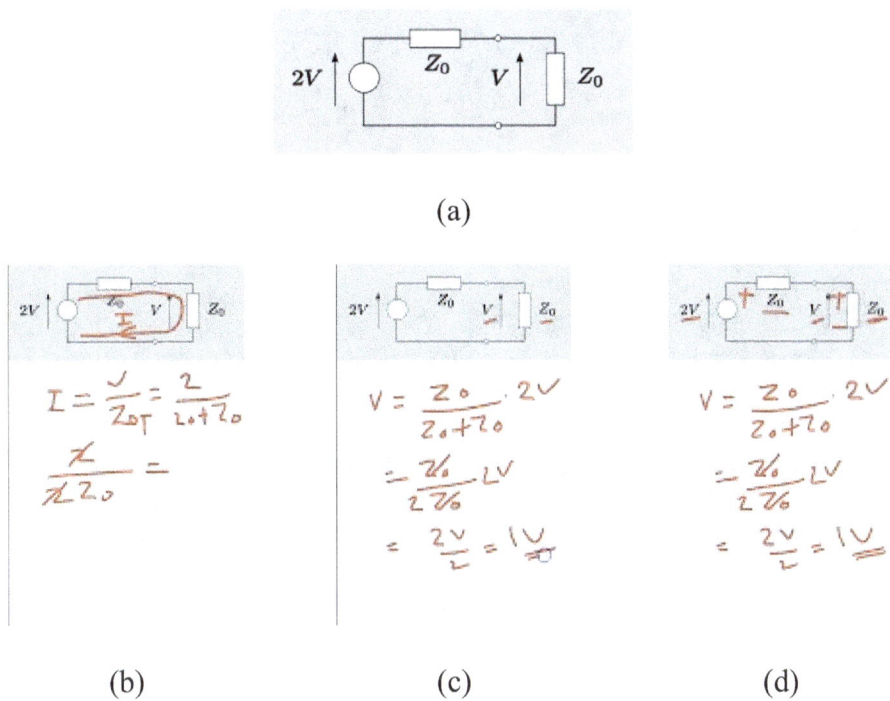

(a)

(b)

$$I = \frac{V}{Z_T} = \frac{2}{Z_0 + Z_0}$$

$$\frac{2}{2Z_0} =$$

(c)

$$V = \frac{Z_0}{Z_0 + Z_0} \cdot 2V$$

$$= \frac{Z_0}{2Z_0} \cdot 2V$$

$$= \frac{2V}{2} = 1V$$

(d)

$$V = \frac{Z_0}{Z_0 + Z_0} \cdot 2V$$

$$= \frac{Z_0}{2Z_0} \cdot 2V$$

$$= \frac{2V}{2} = 1V$$

Figure 3-18 Another example of voltage division [28]

In this circuit (Figure 3-18 Another example of voltage division [28]), we have an impedance Z_0 connected in series. There are two Z_0 impedances in the circuit, and they share the same value. We also have a 2-volt voltage source connected to the circuit. Let's calculate the voltage across Z_0.

Calculating the current (I) flowing through the circuit: According to Ohm's law:

$I = V / Z_{\text{total}}$

$I = 2 \text{ volts} / (2 * Z_0)$

$I = 1 / Z_0 \text{ Amperes}$

If Z_naught is equal to 1 ohm, then the current (I) would be 1 Ampere.

Using the **voltage division** rule to find the voltage (V) across Z_0:

$V = Z_0 / (Z_0 + Z_0) * 2 \text{ volts}$

$V = (Z_0 / 2Z\_naught) * 2 \text{ volts}$

$V = 1 \text{ volt}$

So, the voltage drop across Z_0 is 1 volt.

Applying Kirchhoff's voltage law (KVL): $2 \text{ volts} - I Z_0 - I Z_0 = 0$

By solving this equation, you can determine the values of voltages and currents in the circuit.

Example:

So, let us challenge you through another circuit to calculate the voltage V out as shown in Figure 3-19. Imagine you are given a supply voltage Vin (or input voltage Vin), and there are three resistors: R1, R2, and R3, connected as shown. We see that R2 and R3 are in parallel, and to this parallel combination of R2 and R3, we have R1 in series with that. The total resistance, RT, can be written as R1 + R2 || R3 (i.e. R2 in parallel with R3).

Now, Vout can be calculated using the voltage division rule. Vout can also be calculated by applying Ohm's law. You calculate the total current flowing into the circuit, and then you calculate the voltage drops across each resistor.

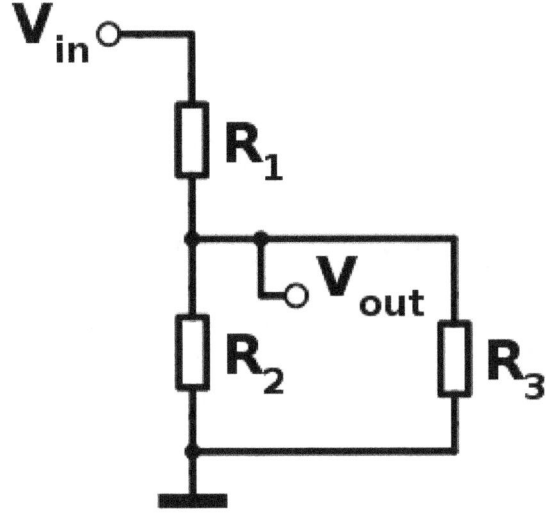

Figure 3-19 currents and voltage calculations [29]

Here's how you can calculate the current (I) that flows through the circuit: I = Vin / RT, where RT is R1 + R2 || R3 (i.e. R2 in parallel with R3).

Now, let's calculate the individual voltage drops:

1. Voltage across R1 is VR1 = I * R1

2. Voltage across R2 is VR2 = IR2

3. Voltage across R3 is VR3 = IR3

These currents get divided into two parts, IR2 and IR3, which, when multiplied by their respective resistors, provide the voltage drops across the individual resistors. Now, for Vout, we need to find the voltage across the parallel combination of R2 and R3, which we can call RX:

RX = (R2 * R3) / (R2 + R3)

Now, Vout can be calculated as follows:

Vout = (RX / (RX + R1)) * Vin

This equation helps you calculate Vout. Using the voltage division rule, you can simplify it by noting that Vout is equal to the voltage across RX, which is the voltage across R2 and R3. So, Vout = VR2 = VR3.

So, in summary, you can calculate V out using the voltage division rule or Ohm's law.

b. Applying current division in circuits

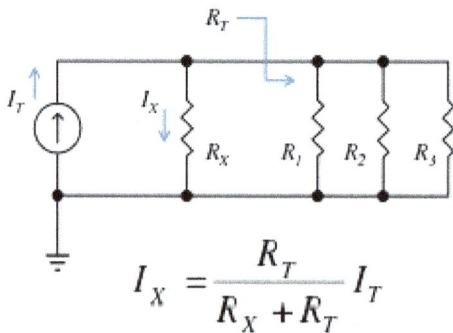

$$I_X = \frac{R_T}{R_X + R_T} I_T$$

Figure 3-20 Schematic of an electrical circuit illustrating current division. Notation RT refers to the total resistance of the circuit to the right of resistor R_X [30].

In this circuit, we will explore the application of the current division rule. The circuit consists of a resistive network with a current source labeled I_T, which represents the total current entering the circuit. The direction of the arrow indicates the current's flow. We have three resistors: R_1, R_2, and R_3, as well as an additional resistor labeled R_X, through which the current I_X flows. The total equivalent resistance of R_1, R_2, and R_3 is represented as R_T.

If we are tasked with calculating the current flowing through R_X (I_X), we can apply the current division rule. The formula for calculating I_X is as follows:

$$I_X = (R_T / (R_T + R_X)) * I_T.$$

Here, R_T represents the equivalent resistance of the three resistors (R_1, R_2, and R_3) connected in parallel. We divide this value by the sum of the total resistance and R_X and then multiply it by the total current, I_T.

Suppose we need to calculate the current flowing through R_3. In that case, we follow a similar approach by first finding the equivalent resistance of R_X, R_1, and R_2, denoted as "RTA." The formula for calculating the current through R_3 (I_{R3}) is:

$I_{R3} = (R_T / (R_3 + R_T)) * I_T$.

To calculate the current through R_2, we find the equivalent resistance R_{TB} of R_X, R_1, and R_3, using a parallel combination. Then, the formula to determine the current through R_2 (I_{R2}) is:

$R_2 = (R_T / (R_2 + R_T)) * I_T$.

This approach demonstrates how to use the current division rule to calculate branch currents in any given circuit.

Chapter 4   Semiconductors

I.   What is Semiconductor?

In this module, we aim to understand the concept of semiconductors. A semiconductor is a material with electrical conductivity lying between that of a conductor and an insulator (Figure 4-1). Conductors, such as copper and iron, exhibit high electrical conductivity, while insulators like glass and rubber have low conductivity.

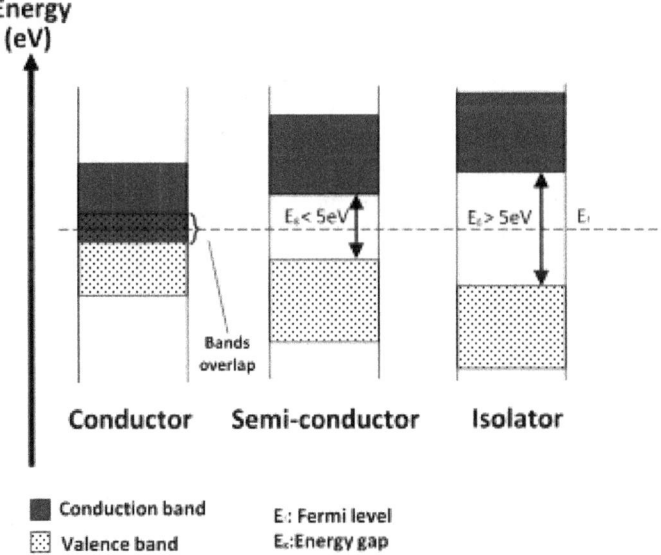

Figure 4-1 Electrons and energy band in materials

Semiconductors, like gallium arsenide, germanium, and silicon, have intermediate conductivity and are crucial for controlling and managing electric current flow in electronic devices and equipment. To comprehend how semiconductors function, let's examine their energy band diagram. The energy band diagram illustrates the relationship between the conduction band and the valence band in different materials. In a conductor, the conduction and valence bands overlap, enabling charge carriers (electrons or holes) to easily transition from the valence band to the conduction band, facilitating electric current flow.

In contrast, insulators have a large energy gap between the valence and conduction bands (greater than 5 electron volts), necessitating a substantial energy input for charge carriers to move from the valence to the conduction band. Consequently, insulators have limited electrical conductivity.

Now, let's focus on semiconductors. Semiconductors possess a smaller energy band gap compared to insulators, making it relatively easier for charge carriers to transition between the valence and conduction bands. This transition can occur through processes such as doping or temperature changes. As a result, semiconductors offer the ability to control and manage electric current flow, making them a fundamental choice in electronic equipment, devices, circuits, and systems.

Let's take a closer look at the energy band diagram of a semiconductor, Figure 4-2. Semiconductors come in two primary types: p-type and n-type semiconductors. When a pure semiconductor is doped with specific impurities, it transforms into an extrinsic semiconductor. Extrinsic semiconductors further divide into two categories: n-type and p-type semiconductors.

In Figure 4-2, we focus on the energy band diagram of an n-type semiconductor. Like any semiconductor, it consists of key energy bands: the dark gray region represents the conduction band, while the white dotted area symbolizes the valence band. However, in a doped semiconductor, such as an n-type semiconductor, there's an additional energy band known as the donor band, indicated by the dashed lines. This donor band signifies the presence of excess electrons, making it an n-type semiconductor, where a surplus of electrons is available for conduction.

II.   Energy band in Semiconductors

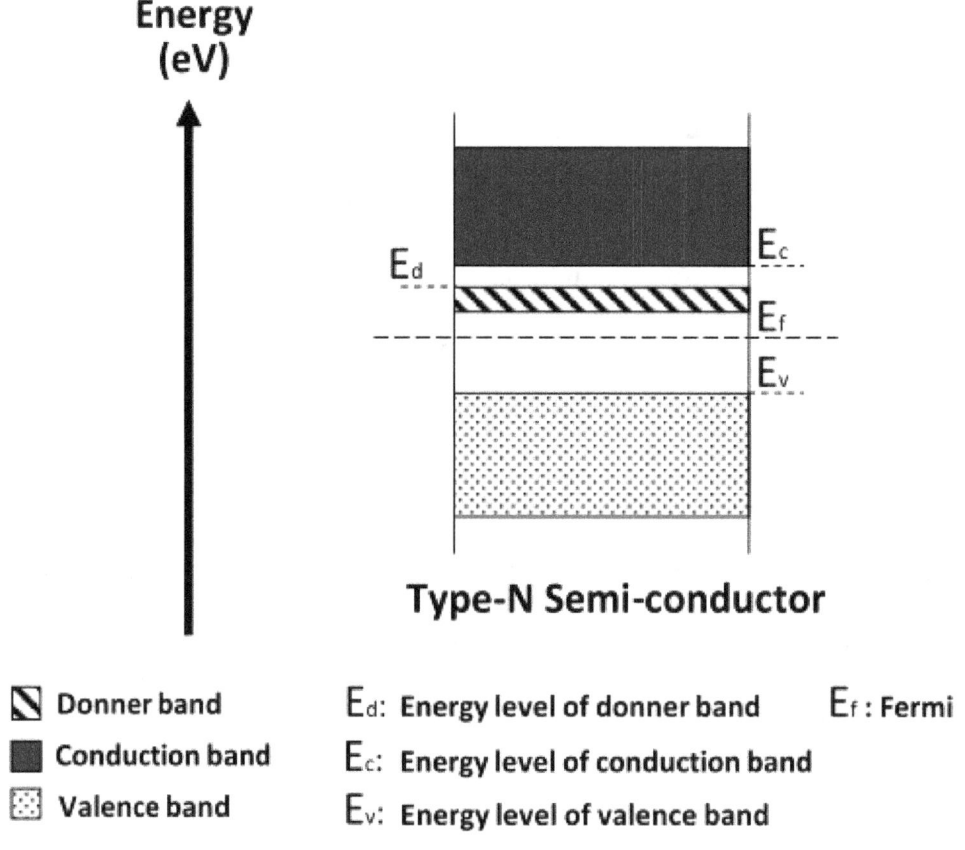

Figure 4-2 Energy band in semiconductors [31]

From an energy level perspective, we have the following designations:

- EC (Conduction Band Energy Level)

- EV (Valence Band Energy Level)

- EF (Fermi Level)

The Fermi level (EF) plays a pivotal role in semiconductor conductivity. It represents the energy level at which an electron must attain sufficient energy to transition from the valence band to the conduction band, enabling the flow of electric current. In n-type semiconductors, the presence of the donor band ensures a surplus of electrons, making it easier for current to flow.

Doping a semiconductor, as depicted in this energy band diagram, signifies that the material has been altered to become an n-type semiconductor, characterized by a substantial number of charge carriers, in this case, electrons, available to facilitate efficient electrical conduction within the material

III.   Doping in Semiconductor

Doping involves the introduction of impurities into a pure semiconductor material like silicon. This process enhances or increases the semiconductor's conductivity. Let's delve into the details:

In its pure form, silicon atoms each possess four valence electrons in their outer orbits. For instance, every silicon atom has four electrons in its outer shell, and they form bonds with neighboring silicon atoms. This bonding is known as covalent bonding, where one electron from one silicon atom pairs with one electron from a neighboring silicon atom, creating a bond. This covalent bonding occurs throughout the crystal lattice structure.

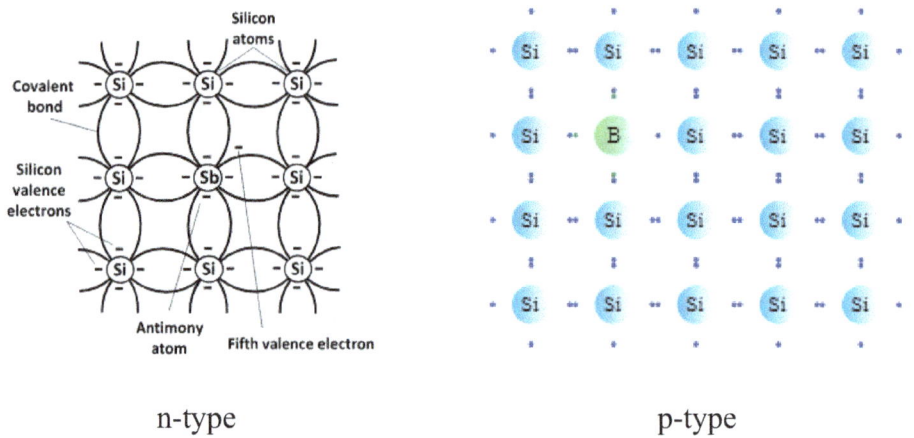

n-type   p-type

Figure 4-3 Silicon doping transforming the pure form into n-type and p-type semiconductors

Now, when we introduce an impurity into the pure silicon, let's say a semiconductor material like Antimony (Sb), which is termed a 'donor,' something interesting happens. Antimony has five electrons in its outer shell, as indicated by the five electrons in Figure 4-3. Four of these electrons form covalent bonds with neighboring silicon atoms, just like in the intrinsic silicon crystal. However, there's an extra or fifth electron left unbound.

This introduction of an impurity is referred to as 'doping.' Due to this doping process, we now have an excess of electrons within the crystal lattice. Consequently, the intrinsic silicon, which was initially a pure semiconductor, transforms into an 'N-type' semiconductor. The 'N' stands for 'negative' because of the excess electrons introduced by the donor impurity.

Mathematically, the excess electron density in the N-type semiconductor can be described as the difference between the number of electrons introduced by the donor impurity (such as phosphorus) and the number of holes (missing electrons) in the intrinsic silicon crystal lattice. This excess electron density, represented by 'n,' plays a crucial role in the conductivity of the N-type semiconductor.

Let's examine another example to understand the process of doping in silicon material. In the second image of Figure 4-3, we can observe silicon atoms, each of which has four valence electrons in their outer orbits. These electrons are referred to as silicon valence electrons. It's important to note that each silicon atom possesses four such outer electrons, establishing the foundation for covalent bonding.

The covalent bonding is illustrated in the diagram, where an electron in the outer shell of each silicon atom forms a bond with a neighboring electron present in the outer shell of an adjacent silicon atom. This bonding pattern is a characteristic feature of pure silicon, which is a pristine semiconductor.

Now, let's introduce an external impurity into this pure silicon material. For this example, we'll consider Boron (B) as the impurity. Boron, unlike silicon, has only three valence electrons in its outer shell. When we add a boron atom (depicted in the image), it contributes its three valence electrons to form covalent bonds with the valence electrons of neighboring silicon atoms. However, there is a deficiency or void of one electron because boron has only three electrons to offer.

This deficiency or absence of an electron creates what is known as a 'hole' within the crystal lattice structure. This hole is essentially a vacant position where an electron should be. As we introduce more boron atoms, more such holes are created in the silicon structure due to the shortage of electrons from the boron impurity.

This process of introducing an impurity, like boron, into the silicon lattice creates numerous holes within the structure. As a result, the material becomes a 'p-type' semiconductor. The 'p' in 'p-type' signifies the prevalence of positively charged holes in the material due to the absence of electrons in these positions.

In summary, the doping process involving boron impurity transforms the pure silicon, which initially had a balanced number of electrons and holes, into a p-type semiconductor characterized by a surplus of positively charged holes.

Doping plays a crucial role in the fabrication of semiconductor devices, as it allows precise control over the electrical behavior of these materials. By strategically introducing impurities, engineers can create N-type and P-type regions within a semiconductor device, enabling the development of diodes, transistors, and other electronic components that form the basis of modern electronics. This control over electrical properties is essential for the design and operation of semiconductor devices.

Here are some of the most commonly used n-donors and p-acceptors:

N-Dopants (Electron Donors):

1. **Phosphorus (P):** Phosphorus is one of the most commonly used n-dopants. It has five valence electrons, and when it is introduced into silicon, for example, it contributes an extra electron, creating an excess of electrons in the crystal lattice.

2. **Arsenic (As):** Arsenic, like phosphorus, has five valence electrons. When doped into silicon, it also acts as an n-dopant by providing an additional electron.

3. **Antimony (Sb):** Antimony is another element with five valence electrons, making it suitable as an n-dopant in semiconductors. It donates an extra electron to the crystal structure.

P-Dopants (Hole Acceptors):

1. **Boron (B):** Boron is one of the most commonly used p-dopants. It has only three valence electrons, and when incorporated into a semiconductor lattice, it creates 'holes' by forming bonds with neighboring atoms, leaving vacancies where electrons should be.

2. **Gallium (Ga):** Gallium is another element used as a p-dopant. It also has three valence electrons and introduces 'holes' into the crystal structure.

3. **Indium (In):** Indium, like boron and gallium, has three valence electrons and can be used to create p-type semiconductors by introducing holes into the material.

IV. PN Junction

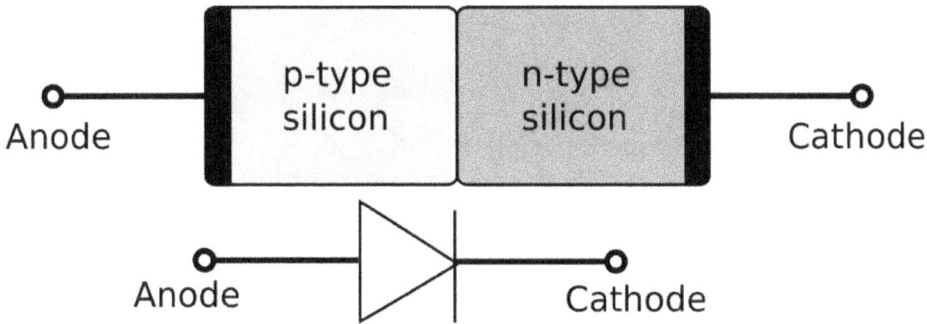

Figure 4-4 PN junction and diode symbol [32]

In Figure 4-4, we observe the concept of P-N junction formation. When you combine a p-type semiconductor material with an n-type semiconductor material, you create what is known as a PN Junction. This fundamental junction forms the basis of various semiconductor devices, including diodes and transistors.

A PN junction is a crucial component in semiconductor electronics. It's formed by joining a region of p-type material, where there's an excess of 'holes' (positive charge carriers due to missing electrons), with a region of n-type material, characterized by an excess of electrons. At the boundary where these two types of materials meet, interesting electronic properties emerge.

This junction allows for the controlled flow of charge carriers, depending on the voltage applied. In a forward bias, it facilitates the flow of current, while in a reverse bias, it acts as a barrier to current flow. This property of the PN junction is harnessed in diodes to allow current flow in one direction only, making them useful for rectification and signal switching. Transistors, on the other hand, leverage the PN junction's ability to amplify and control current for various electronic applications. This junction plays a pivotal role in various semiconductor devices due to its unique electronic properties.

a. PN junction semiconductor diodes

Referring to Figure 4-4, we are looking at two types of silicon materials: p-type silicon and n-type silicon. Both of these materials belong to the category of semiconductors. P-type semiconductor is created by introducing an acceptor-type impurity, while n-type silicon is formed by adding a donor-type impurity to the silicon material. Both p-type and n-type silicon are examples of extrinsic semiconductors, achieved through a process called doping.

Now, at one end of the p-type silicon, we have a terminal referred to as the 'anode,' as indicated here. At the other end of the n-type silicon semiconductor, we have a terminal known as the 'cathode.' When these two types of semiconductors are joined together, the region where they meet is called the 'junction.' This combination gives rise to what we call a 'PN junction.' The entire device is commonly known as a 'PN Junction Diode.'

In Figure 4-4, you can see the symbol representing a PN junction diode, where one terminal is designated as the 'anode,' and the other terminal is labeled as the 'cathode.' It's essential to remember this symbol when working with PN junction diodes, as it is widely used in electronic schematics and diagrams.

b. Physics of PN junction semiconductor devices

In this module, we aim to understand the functioning of a PN junction (Figure 4-5). Here, you can see a P-type semiconductor referred to as P (with an abundance of holes) and an N-type semiconductor (with a surplus of electrons). P-type semiconductors predominantly contain holes as their majority charge carriers, while N-type semiconductors are rich in electrons as the majority charge carriers.

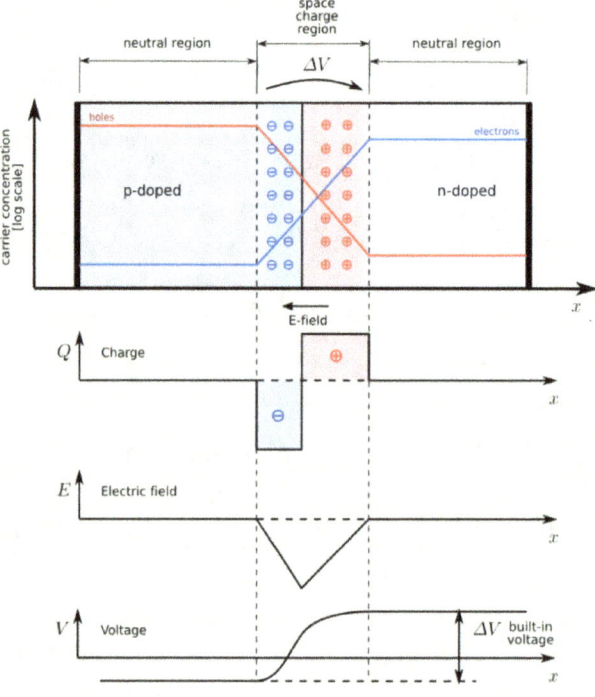

Figure 4-5 Physics of PN junction semiconductor devices

When a P-type semiconductor is joined with an N-type semiconductor, as indicated by the line here, electrons (majority carriers) and holes (positive charge carriers, majority in P-type) attempt to diffuse from one region to another due to electrostatic forces. This phenomenon is known as the diffusion process. Electrons from the N-type region migrate to the P-type region and fill some of the holes (shown in blue). Conversely, some holes travel from the P-type region to the N-type region and fill with electrons (shown in blue). This intermingling of charge carriers occurs within the space charge region or depletion region.

The depletion region forms a built-in potential or built-in voltage across the PN junction, denoted as ΔV. This voltage arises due to the presence of positive ions (red symbols) and negative ions (blue symbols) within this region. It's important to note that the depletion region lacks free electrons or holes; instead, it contains these ions.

Next, let's discuss some electrical quantities concerning the charge distribution along the x-axis. In the depletion region (the blue zone with negative ions), the charge reaches its maximum. At the junction, the charge is zero, and it increases negatively in the depletion region where negative ions are present. Conversely, in the red region where positive ions exist, the charge increases positively. This charge distribution defines the electric field. The voltage profile shown here demonstrates that there is a voltage in the depletion region (ΔV), which is the built-in voltage. To facilitate the flow of electricity in this PN junction, an external voltage is applied across it. By means of this applied electric field, we need to overcome the depletion region voltage, which is the barrier voltage or built-in voltage. This allows electrons (or holes) to traverse from one region to another, effectively completing the circuit.

In essence, when two semiconductor materials are joined to form a PN junction, the application of an external voltage enables the flow of current. By overcoming the internal barrier voltage, the electric field facilitates the movement of electrons (or holes) from one region to another, resulting in conduction.

Chapter 5   Semiconductor Devices

I.   Diode

a.   PN junction silicon and germanium diode

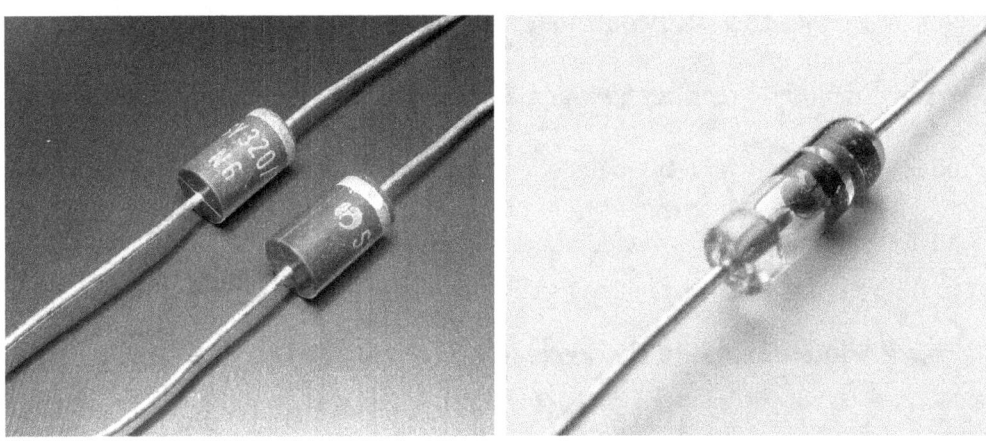

Silicon diode [33]   Germanium diode [34]

Figure 5-1 real diodes

Figure 5-1 shows real PN Junction silicon and germanium diodes, with identifiable terminals. The end with the silver ring (silicon) or black ring (germanium) is referred to as the cathode, while the opposite terminal is called the anode. It's important to remember that the symbol for this diode is as shown in Figure 5-1.

Silicon diodes typically have a cut-in voltage of around 0.7 volts for standard silicon diodes. This means that they begin to conduct current when a forward voltage of approximately 0.7V is applied across them. Germanium diodes have a lower cut-in voltage compared to silicon diodes. They typically have a cut-in voltage of around 0.3 to 0.4 volts. This makes them suitable for low-voltage applications. Silicon diodes are widely used in various electronic circuits, including rectifiers, voltage clippers, and signal clamps. Germanium diodes have been largely replaced by silicon diodes in modern electronics due to their lower performance at higher temperatures.

b. PN junction diode IV current-voltage curve

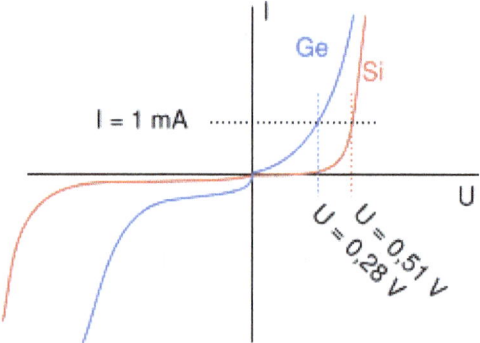

Figure 5-2 PN junction diode input-output characteristics [35]

In this segment, we observe the typical IV (current-voltage) characteristics of a PN Junction diode (Figure 5-2). The IV characteristics depict the relationship between current (I) and voltage (V). There are two common semiconductor diodes: one made of germanium and the other of silicon. The red curve represents the IV curve for the silicon diode, while the blue curve represents the IV curve for the germanium diode.

When you forward bias a PN Junction diode, it means you apply a positive voltage to the P-type material and a negative voltage to the N-type material. This action moves the diode into forward bias, as illustrated. The cut-in voltage, also known as the threshold voltage, is approximately 0.5 volts for silicon diodes. At this point, the diode begins to conduct and acts as a short circuit. Current flows from P to N, and the circuit is closed. Until the voltage reaches 0.5 volts on the x-axis, the current in the diode remains minimal. The same behavior applies to germanium diodes, but their cut-in voltage is approximately 0.2 volts. Beyond this voltage, the current starts to increase.

Now, let's discuss the reverse bias characteristics. When the diode is reverse-biased, as indicated in the diagram, the current through the diode remains very small as the voltage is increased. This is due to the fact that the junction is reverse-biased and effectively acts as an open switch. However, as you continue to increase the voltage, there is a sudden breakdown, and heavy current begins to flow through the diode. This is known as the breakdown region, and it is experienced by both silicon and germanium diodes. Understanding these IV characteristics is crucial for using diodes effectively in electronic circuits.

c. Region of operations of pn junction semiconductor diode | I-V curve

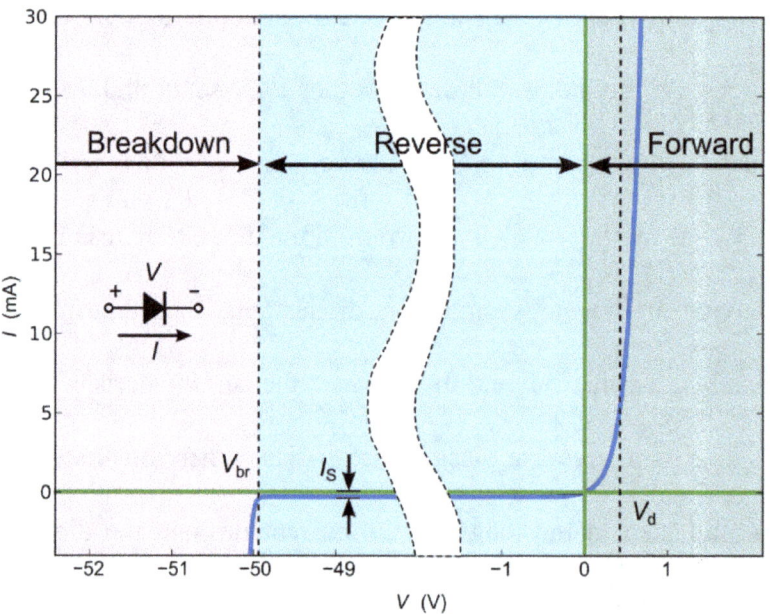

Figure 5-3 Diode regions of operation [36]

1. Forward Bias Region: When a positive voltage is applied to the diode (forward bias), starting from zero voltage and gradually increasing,

there is initially a small current flow. However, as the applied voltage reaches the diode's threshold voltage, often referred to as the "cut-in" voltage (VD), there is a sudden exponential increase in the diode's current. This region, represented by the green curve, signifies the forward biasing of the diode. In this region, the diode is conducting current efficiently.

2. Reverse Bias Region: Conversely, when a negative voltage is applied across the PN Junction diode (reverse bias), starting from zero and increasing in the negative direction, the diode exhibits minimal or ideally zero current flow. This region, shown by the blue curve, indicates reverse bias operation. In this situation, only a negligible amount of current, known as reverse leakage current, flows through the diode.

3. Breakdown Region: In the breakdown region, as the reverse voltage continues to increase, there comes a point where the diode suddenly exhibits a significant increase in current. This phenomenon is known as breakdown. When the reverse voltage reaches a certain critical value, known as the breakdown voltage, or reverse breakdown voltage

(VBR), the diode experiences an avalanche effect, and the current surges rapidly. Operating the diode in this region can lead to damage.

These three regions are the fundamental operating characteristics of a PN Junction diode. Understanding and properly utilizing these characteristics are essential for designing and using electronic circuits involving diodes.

II.   **Bipolar junction transistor**

Figure 5-4 How BJT looks like [37]

In Figure 5-4, we observe a transistor, specifically a BJT (Bipolar Junction Transistor). A BJT is a semiconductor device, and it derives its name from the fact that it conducts electricity using two types of charge carriers:

electrons and holes. Depending on the type of charge carriers it utilizes for conduction, it can be classified as either an NPN or PNP transistor.

There are two main types of transistors:

1. NPN Transistor: This type of transistor employs electrons to carry the current.

2. PNP Transistor: PNP transistors use holes to conduct the current.

Transistors typically feature three terminals:

1. Emitter

2. Base

3. Collector

These components are found in various forms, including integrated circuits (ICs), metal cans, and ceramic packages. That concludes our overview of transistors. We have discussed transistors in detail in the following sections.

III.   Transistor types and configurations

a. Bipolar junction transistor (BJT) operation

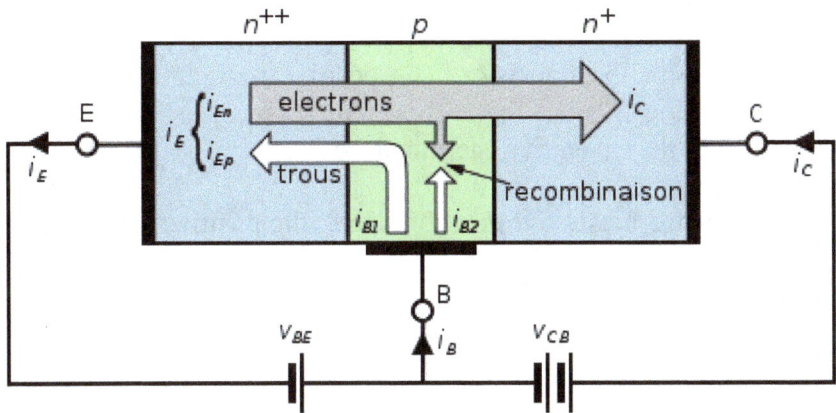

Figure 5-5 Bipolar junction transistor operation and biasing [38]

The diagram in Figure 5-5 illustrates the operation of an NPN transistor. In this configuration, 'N' represents the N-type semiconductor, while 'P' represents the P-type semiconductor. The transistor consists of three layers, with a P-layer sandwiched between two N-type layers, hence the term 'NPN transistor.'

Let's examine the voltage sources connected to the transistor. The first voltage source, denoted as 'VBE,' is connected between the base and emitter. The positive terminal of VBE is linked to the base, while the negative terminal connects to the emitter. This arrangement forward-biases the PN junction, with the positive side connected to the P-type semiconductor and the negative side to the N-type semiconductor.

Now, let's focus on the other voltage source, 'VCB,' connected across the collector and base of the transistor. The positive terminal is linked to the collector, while the negative terminal connects to the base. This configuration reverse-biases the junction. Here, the positive side connects to the N-type material, and the negative side connects to the P-type semiconductor.

This combination results in one junction being forward-biased, while the other is reverse-biased. This is the fundamental condition for a transistor to operate in its active region.

Turning our attention to the circuit, we have the collector current, denoted as IC, flowing into the collector. Simultaneously, the emitter current, IE, flows out. Additionally, there is a base current, IB, flowing into the base region. Mathematically, IE is the sum of IB and IC.

From a physics perspective, the direction of electron flow is opposite to conventional current flow. Electrons move from the emitter towards the collector. Some of these electrons recombine with the holes in the base region, signifying recombination. This figure effectively explains the operation of an NPN transistor when biased in its active region.

b. BJT and its symbols

An NPN transistor consists of three semiconductor layers: N-type, P-type, and N-type, arranged in that order. Referring to Figure 5-6, the arrow in the symbol represents the direction of conventional current flow, which goes from the collector to the emitter, with little to no current flowing into the base in an ideal transistor. Electrons, which are negatively charged, flow in the opposite direction, from emitter to collector, to allow this conventional current flow.

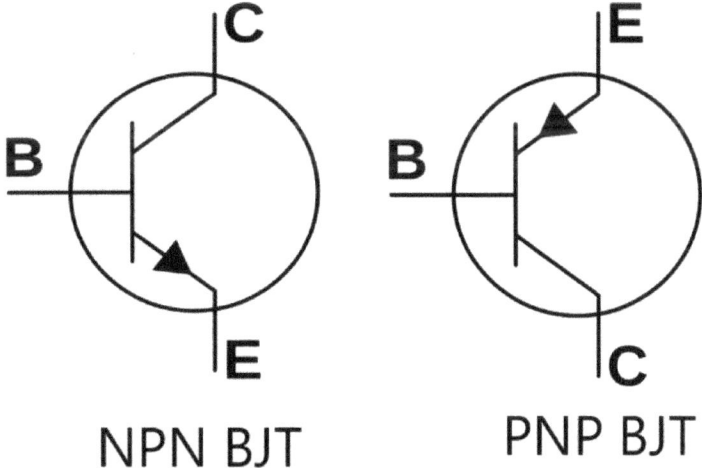

NPN BJT   PNP BJT

Figure 5-6 BJT symbols [39]

In case of a PNP transistor, P stands for P-type semiconductor and N stands for N-type semiconductor. In this configuration, a P-type semiconductor material is sandwiched between two adjacent N-type semiconductor materials. Referring to Figure 5-6, the arrow in the symbol still represents the direction of conventional current flow, which goes from the emitter to the collector in a PNP transistor, with little to no current flowing into the base in an ideal transistor. Electrons, which are negatively charged, flow in the opposite direction, from collector to emitter, to allow this conventional current flow.

c. Transistor as a switch and amplifier

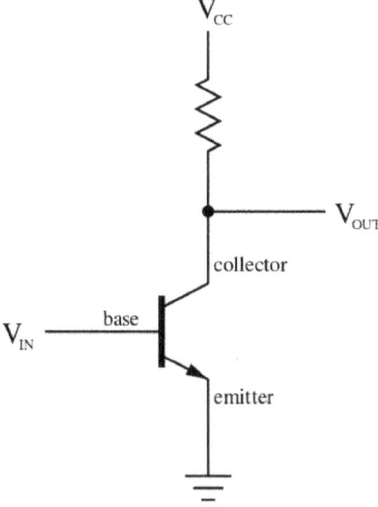

Figure 5-7 Transistor as a switch/amplifier [40]

Let's delve into the operation of a transistor through this simple circuit (Figure 5-7). As you can see, it's an NPN transistor, characterized by the direction of current flow. In this configuration, the N-region is sandwiched between two P-regions, and the current flows from the collector to the emitter, hence the name NPN transistor. Now, when an input signal is applied to the base of the transistor, the base-emitter junction becomes forward-biased. This results in a small current called the base current (IB) flowing into the input side of the circuit. The transistor's gain, denoted by β (beta), amplifies this base current to produce the collector current (IC) at the output side of the circuit. In essence, the small input signal is transformed into a larger output signal, illustrating the transistor's amplification capability.

This same circuit, aside from acting as an amplifier, can also function as a switch. Imagine when the input signal is at zero volts; this effectively turns off the input side of the circuit, causing no input current, and consequently, no output current. Therefore, the output voltage (V_{out}) is tied to VCC.

Conversely, when a high voltage signal is applied, it forward-biases the base-emitter junction, allowing a base current (I_B) to flow, which, when amplified by the transistor, results in an output current (I_C). This current passing through the resistor (R) causes a voltage drop, so the output voltage (V_{out}) becomes V_{CC} - I_C * R, which is lower than V_{CC}. This behavior effectively demonstrates how the transistor can function as a switch.

d. Example

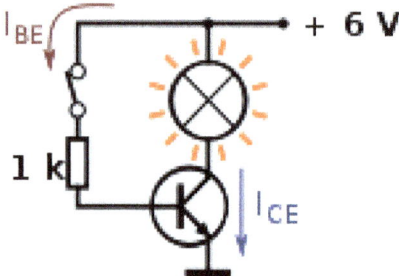

Figure 5-8 Transistor as a switch

In this circuit (Figure 5-8), we have an NPN transistor with distinct collector, base, and emitter terminals. The emitter is grounded, the collector is linked to a 6-volt supply, and the base is under the control of a switch.

When the switch is in the open position, it disrupts the current flow through the resistor since the circuit is incomplete. Consequently, no current enters the input side of the transistor. This leads to the transistor being in an "off" state because there is no current on the input side. Therefore, the transistor acts as an open circuit, and the light bulb connected to it remains unlit.

However, when the switch is closed, it closes the circuit, enabling a current (IB) to flow through the resistor and into the input side of the transistor. We can apply Kirchhoff's voltage law to the input side of the transistor, which can be expressed as follows:

$6 \text{ V} - I_B * 1 \text{ k}\Omega - V_{BE} = 0.$

Here, V_{BE} represents the forward voltage drop across the base-emitter junction of the transistor, typically around 0.6 to 0.7 volts for a silicon transistor. As a result, the voltage across the input side decreases.

Considering the forward voltage drop, I_B induces a corresponding current, known as the collector current (I_C), to flow from the collector to the emitter on the output side of the transistor. This establishes a conduction path from the supply voltage to ground, and the light bulb connected in this path will illuminate.

In summary, when the switch is closed, the transistor enters the "on" state, permitting current to flow from the collector to the emitter, and the bulb lights up. Conversely, when the switch is open, the transistor is in the "off" state, interrupting the current flow and keeping the bulb off. This illustrates how a transistor functions as a switch, regulating current flow in an electronic circuit.

IV.   Field effect transistor (FET)

a. How does a Field Effect Transistor works? | FET vs BJT

This current is controlled by an input voltage applied at the gate. This is why it's called a Field Effect Transistor, as the applied electric field at the gate controls the current in this device.

Now, let's examine the structure of this semiconductor device. It consists of a P-type semiconductor material as shown, known as the body of the FET, with two N+ doped regions within it. The N+ regions have plenty of electrons, while the P-type material has plenty of holes. The P-type material contains holes as the majority carriers and electrons as minority carriers, whereas the N-type material contains electrons as the majority carriers and holes as minority

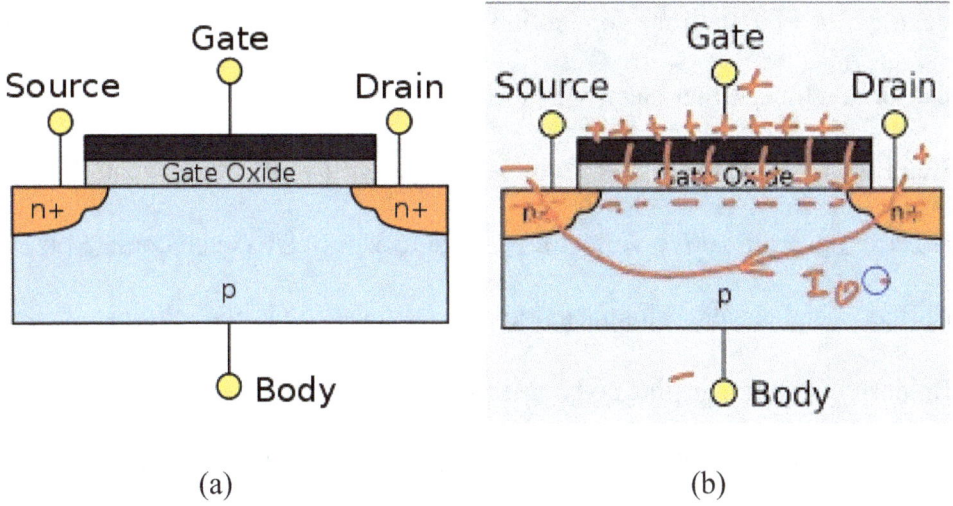

(a)   (b)

Figure 5-9 Field effect transistor [41]

carriers. In total, there are four pins for the device: body (a P-type substrate),

gate, source, and drain. On the body of the P-type semiconductor, an

insulator material is deposited, known as gate oxide. Typically, this material

is SiO2 (silicon dioxide), although silicon nitride can also be used. On top of

the insulator, there is a layer of metal or polysilicon, which is why it's also

called a Metal-Oxide-Semiconductor or MOS. In specific cases, it's referred

to as a MOSFET (Metal-Oxide-Semiconductor Field Effect Transistor).

Now, there are two types of FETs: MOSFET and Junction Field Effect

Transistor (JFET). JFET is an example of a unipolar device, unlike the

Bipolar Junction Transistor (BJT), which is bipolar. In a BJT, current

conduction involves both types of charge carriers, electrons, and holes. In contrast, in FET, current conduction occurs due to only one type of charge carrier at a time, either electrons or holes.

Referring to the (b) part of (Figure 5-9), when you apply a voltage at the gate (for example, a positive voltage), positive charge carriers accumulate on top of the metal layer. This accumulation causes negative charge carriers (minority carriers) in the body to move closer to the oxide-semiconductor interface. Once a sufficient positive voltage is applied to overcome the threshold voltage of the transistor, enough negative charge carriers accumulate at the oxide-semiconductor interface, allowing an electric field to form across the drain and source regions. This results in current flowing from the drain to the source, completing the circuit. Removing the gate voltage stops current conduction, and reapplying it allows current to flow again, provided there is a voltage across the drain and source.

In summary, the voltage applied at the gate controls the current in the channel region, which is called the channel of the FET. That's the basic concept of Field Effect Transistors.

b. How Field effect transistor controls current | FET full operation

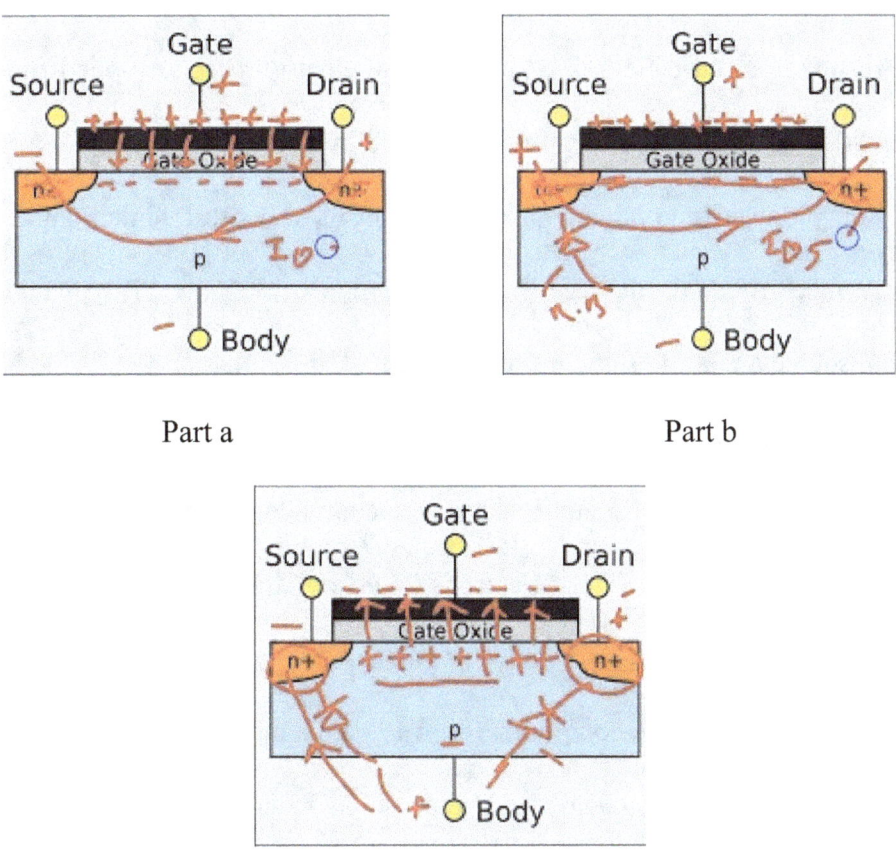

Part a   Part b

Part c

Figure 5-10 Field effect transistor

Let's analyze Field Effect Transistors (FETs) (Refer Figure 5-10, part a).

We've seen that when we apply a positive voltage relative to the body (which

is tied to the negative terminal) at the gate, it deposits positive charges on top

of the gate, which is on top of the metal. If this positive voltage at the gate is sufficient, it will attract electrons from the P-type body towards the gate oxide and the P-type semiconductor interface.

As a result, an electric field is created, and an electric force is applied from the positive charges to the negative charges, following Coulomb's law. Now, if we apply a positive voltage to the drain pin of the transistor while tying the source to the ground or the negative, (part a), remember that the P-type body and the source are at the same negative potential, this situation floods the N-type semiconductor, and the N+ regions at both ends of the transistor, with electrons. In the channel, many electrons accumulate, creating a complete path from source to drain. Thus, electrons flow from drain to source, and this current is called I_{DS} (drain-to-source current).

Now, let's consider diode conditions (refer to part b in the figure). A P-type semiconductor is connected to the N-type semiconductor, with the N-side tied to the positive and the P-side to the negative. This diode is reverse-biased. Similarly, another diode's terminals are also connected to the negative voltages at both the ends. This results in a completely reverse-biased diode. Consequently, the junction diodes do not become forward-biased, which is a favorable situation. Forward-biased diodes in this configuration could interfere with the operation of the FET.

Moving on to the third situation (part c), let's say we apply a negative potential to the gate, with the body tied to the positive potential, creating opposite polarity conditions. This situation causes many negative charge carriers to accumulate on top of the gate, attracting many positive charge carriers to the gate oxide and the P-type semiconductor interface. This creates an electric field, with the direction of electric field lines and electrostatic force (direction of arrows) as shown.

Now, will current flow when you apply a voltage difference between source and drain?

In this case, the N+ regions on either end of the transistor have many electrons, but in the middle, the channel has many holes, which are positive charge carriers. Thus, there is no current flow from source to drain or drain to source.

Regarding the diodes, the P-body is tied to the positive terminal and the N-side to the negative, resulting in a forward-biased condition for this diode. Similarly, for another diode, both P-body and N+ region are connected to the positive voltages. This diode is not forward-biased. Essentially, for part c, no current will flow through the transistor even if you reverse the polarity of source and drain and only one of the junctions are forward-biased.

In summary, we've examined four conditions: gate positive relative to the body, gate negative relative to the body, drain positive relative to source, and source positive relative to drain. These biasing conditions help us understand the current flow and switching action of the FET.

c. Depletion region | n-channel FET | circuit

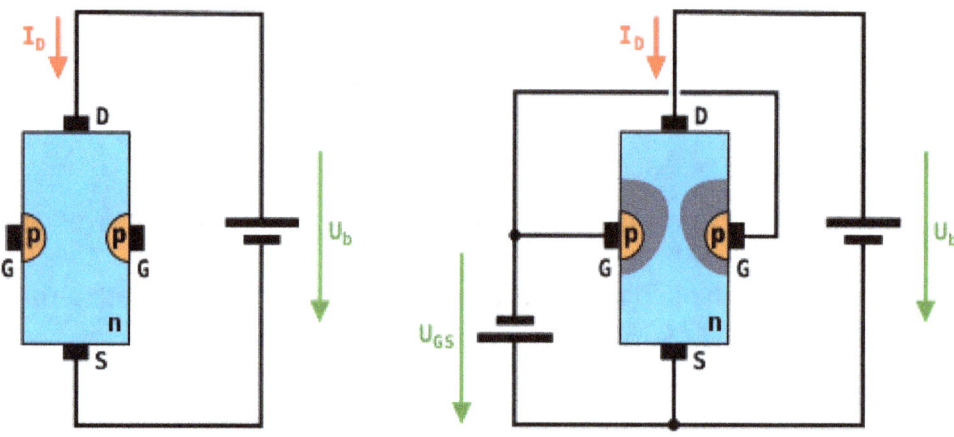

Figure 5-11 How voltage at the Gate controls the depletion region width in n-channel FET [42]

In the provided circuit description, we have a Field Effect Transistor (FET) configuration (Figure 5-11)

To the left, there's an N-type silicon semiconductor bar, and within this N-type material, there are two P-regions, resulting in the formation of two P-N junctions.

On either side of these P-regions, there are two gate pins, which are shorted together and considered as a common gate. Additionally, there are two more pins labeled as source and drain, totaling three pins for this FET.

A voltage source is applied across the drain and source, with the positive terminal connected to the drain and the negative terminal connected to the source. In this setup, the N-type semiconductor, represented by the blue region, acts as a conductive path through which the voltage source supplies current. The current flows from drain to source, as indicated by the arrow's direction. It's important to note that due to the presence of two P-N junctions, depletion regions are formed within the FET's channel, which corresponds to the blue region referred to as the channel of the transistor.

This configuration represents an N-channel FET because current flows through the N-type material.

Now, let's examine the circuit on the right. Here, a battery is connected to the gate terminal with respect to the source terminal. The negative terminal of the battery is connected to the gate of the transistor, while the positive terminal of the battery is connected to the source.

In this setup, a negative voltage is applied to the P-type semiconductor on one side (gate-source region), and a positive voltage is applied to the drain on the other side. Consequently, the P-N junction between the drain and the gate becomes reverse-biased, resulting in the depletion region indicated in the diagram.

Similarly, the source terminal of the N-type material is connected to the negative potential, and the gate terminal is connected to the negative potential as well, or alternatively, the source is connected to the positive potential, and the gate is connected to the negative potential. This setup also results in a reverse-biased P-N junction and the corresponding depletion region.

The magnitude of the negative voltage applied to the gate determines the size of the depletion region. As the negative voltage increases, the depletion region expands, reducing the current flowing from drain to source. Conversely, applying a positive voltage to the gate with respect to the source narrows the depletion region, allowing a larger current to flow from drain to source.

In summary, the gate-source voltage controls the current flowing through the FET's channel. By adjusting this voltage, the current can be effectively controlled. Figure 5-12 shows JFET symbols for both the n-channel and p-channel.

Figure 5-12 JFET symbols

V.   Metal Oxide Semiconductor Field effect transistor (MOSFET)

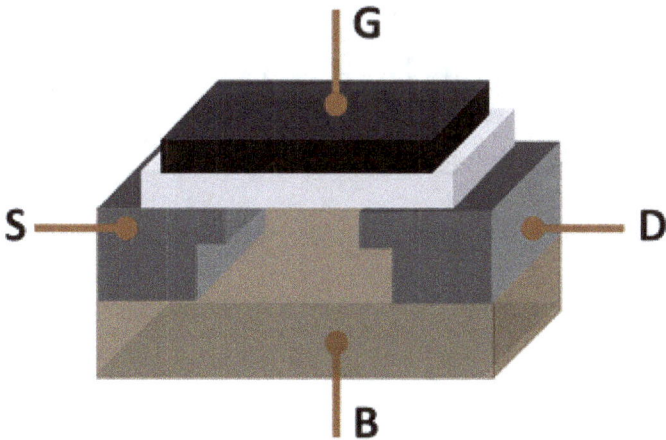

Figure 5-13 MOSFET, showing gate (G), body (B), source (S), and drain (D) terminals. The gate is separated from the body by an insulating layer (light pink) [43].

Let's dive into the details of the basics of MOSFETs.

What is a MOSFET? MOSFET stands for Metal Oxide Semiconductor Field Effect Transistor. In Figure 5-13, you can see the components of a MOSFET. It consists of a silicon substrate, also referred to as the body of the transistor or device. On top of this substrate, there are two terminals: one is called the drain, and the other is the source. Separating these terminals is an insulating layer, typically silicon dioxide or silicon nitride, and above this insulating layer, there is a metal or polysilicon layer. This combination of metal and oxide constitutes the semiconductor. MOSFETs are known for their switching capabilities. For a comprehensive understanding of their operation, we highly recommend checking out our course on CMOS VLSI titled "CMOS VLSI by Dr. Vinayak Pachkawade".

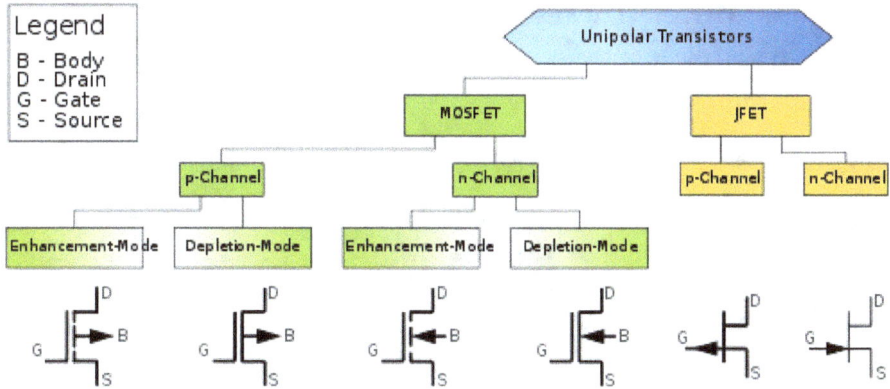

Figure 5-14 FET symbols [44]

MOSFET Symbol The MOSFET symbol represents an extension or modification of the Junction Field Effect Transistor (JFET). MOSFETs operate by applying an electric field, hence the term "Field Effect Transistor." There are two types of MOSFETs: P-channel and N-channel (Figure 5-14). The symbols for these are as follows:

- P-Channel MOSFET: The arrow points outward.

- N-Channel MOSFET: The arrow points inward.

Each MOSFET type has three terminals: Drain (D), Gate (G), and Source (S). Additionally, there is a fourth terminal called the Body (B).

a. How does MOSFET transistor look?

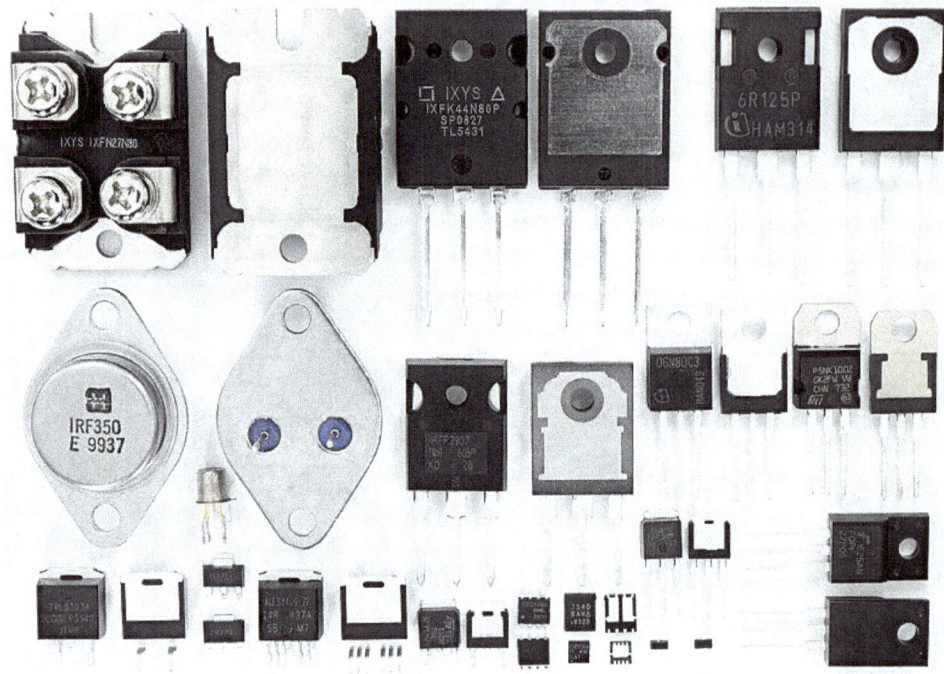

Figure 5-15 MOSFET Transistors [45]

We see a transistor known as a MOSFET, which stands for Metal-Oxide-Semiconductor Field-Effect Transistor Figure 5-15. This is a three-terminal device and falls under the category of discrete electronics semiconductor devices. As you can observe, there are various variations and versions of these transistors. A MOSFET comprises three terminals: the gate, the drain, and the source. To understand which terminal serves what purpose, you can refer to the datasheet specific to the particular transistor. You can locate the code on the body of the transistor, often under a microscope, and enter it into Google to retrieve the relevant datasheet, providing comprehensive information about that MOSFET. Typically, MOSFETs can be classified as N-channel or P-channel, and they come in two primary types: enhancement-type MOSFETs, which are typically off transistors, and depletion-type MOSFETs, which are usually on transistors. In the case of enhancement-type transistors, applying a gate voltage causes an increase in current. This effectively changes the transistor's state from off to on. On the other hand, for depletion-type transistors, the transistor is normally in the on state. By applying a voltage to the gate, you can gradually turn it off by reducing the current level. This provides a fundamental understanding of MOSFETs.

b. Understanding MOSFETs

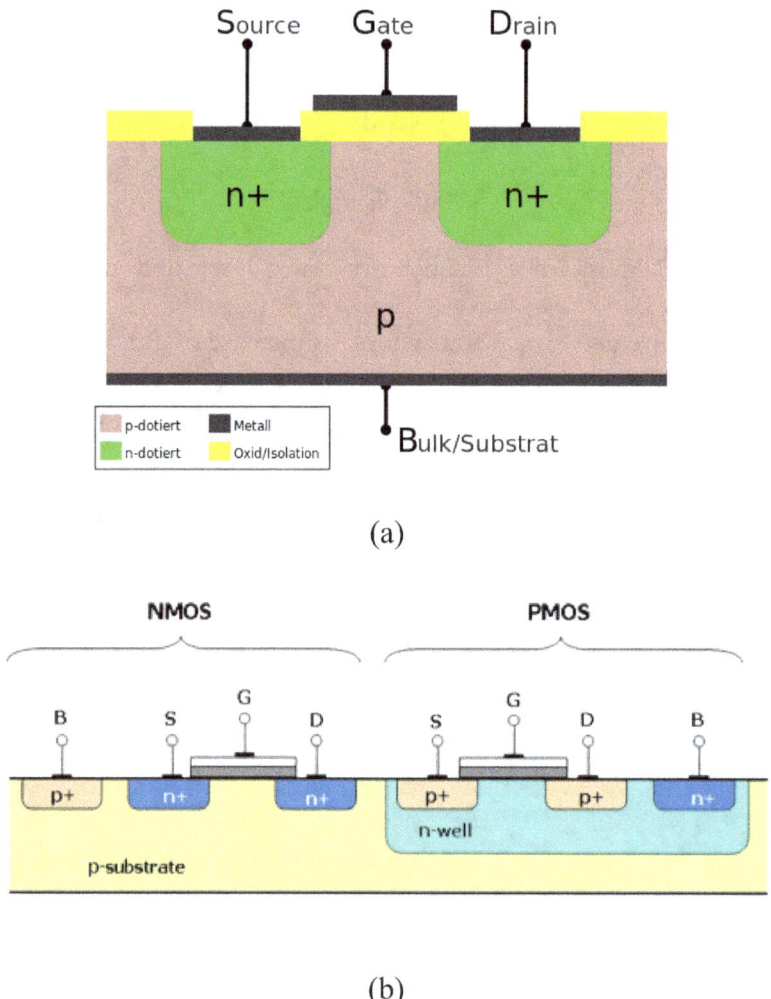

(a)

(b)

Figure 5-16 Scheme of metal oxide semiconductor field-effect transistor (a) [46], Cross section of two transistors in a CMOS gate, in an N-well CMOS process (b) [47]

Let's dive into the world of MOSFETs to gain a deeper understanding. MOSFETs are majority carrier devices, which means that the current flowing through them is primarily due to either electrons or holes. To comprehend their operation, let's examine the structure of a MOSFET.

In Figure 5-16 (a), we have a P-type semiconductor, serving as the body or substrate of the device. Inside this substrate, we find two heavily doped regions labeled as N+ and N+, forming the source and drain, respectively. Separating these regions is an insulating layer, typically made of silicon dioxide or silicon nitride. On top of this insulating layer, we have a metal or polysilicon layer, which acts as the gate.

The MOSFET, with its distinct structure, is capable of controlling current flow from drain to source when a voltage is applied to the gate terminal. It's important to note that MOSFETs come in two fundamental types: N-channel and P-channel.

N-Channel MOSFET (NMOS): In an N-channel MOSFET (NMOS), the majority carriers responsible for current conduction are electrons. When a positive voltage is applied to the gate, a sufficient number of electrons gathers in the channel between the source and the drain. With the drain at a positive terminal and the source at a negative terminal, current flows from source to drain. In other words, electrons move from the source to the drain. The voltage applied at the gate controls this current.

P-Channel MOSFET (PMOS): Conversely, in a P-channel MOSFET (PMOS), the majority carriers in the channel are holes. To create a PMOS transistor, a P substrate is used, with two P+ regions forming the source and drain. Like the NMOS, there is an insulating layer and a gate terminal on top. When an appropriate voltage is applied to the gate, holes in the channel between source and drain enable current flow from drain to source. This current, too, is controlled by the gate voltage.

In MOSFETs, N-channel or P-channel, operate as majority carrier devices. NMOS relies on electrons as majority carriers, while PMOS employs holes. The voltage applied at the gate governs the flow of current in the conducting channel between the source and the drain.

Understanding MOSFETs is essential, as these devices play a pivotal role in modern electronics. Stay tuned for further insights into their operation and applications.

Chapter 6   Diode and Transistor Circuits

I.   Diode circuits

a.   Diode half wave rectifier | AC to DC conversion

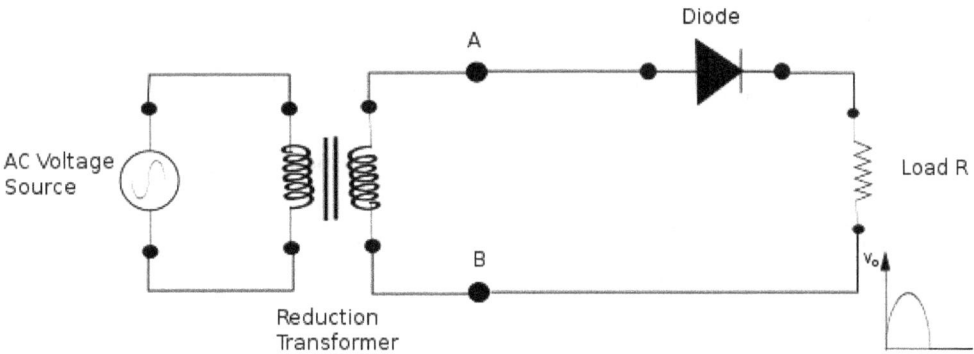

Figure 6-1 Diode half-wave rectifier - AC to DC converter [48]

Let's examine how this circuit operates (Figure 6-1). In this configuration, there is an alternating current (AC) voltage source connected to the primary winding of the transformer. Across the secondary winding of the transformer, we obtain a reduced AC voltage, which is why this transformer is referred to as a step-down transformer. The primary function of this transformer is to convert a high-magnitude AC voltage on the primary side into a lower-magnitude AC voltage on the secondary side.

The secondary voltage, present across points A and B, is sinusoidal in nature. This means it exhibits both positive and negative half cycles.

During the positive half cycle of the alternating signal, the diode becomes forward-biased and functions as a closed switch. Consequently, the positive signal flows through the resistor, and this is reflected in the output waveform, where the positive half cycle is evident.

However, during the negative half cycle of the AC voltage, the diode becomes reverse-biased, effectively acting as an open circuit. Consequently, the signal cannot pass through the diode, and as a result, it does not reach the output load resistors. This omission of the negative half cycle from the output waveform characterizes the operation of this circuit.

When the positive cycle repeats, the diode once again becomes forward-biased, allowing the signal to pass through the load resistance. Consequently, this signal is reflected in the output waveform.

In summary, this circuit functions as a half-wave rectifier or an AC-to-DC voltage converter. It converts a sinusoidal AC waveform into a pulsating DC waveform where the positive half cycles are preserved, but the negative half cycles are omitted. However, it's important to note that this output is not a pure DC voltage, as it exhibits variations over time. For a constant voltage with minimal fluctuations, additional components or circuits are required.

b. Full wave rectifier | AC to DC conversion | diode circuits

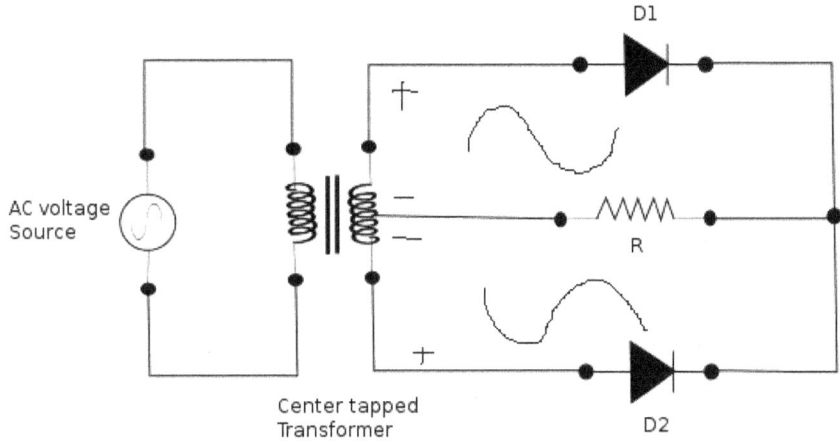

Figure 6-2 Full-wave rectifier - center tap [49]

In this circuit diagram, we will explore how the circuit in Figure 6-2 operates. You have an AC voltage source depicted in the figure. The transformer shown here is known as a Center Tap Transformer because it has a center tap in its secondary winding. As a result, the voltage developed will have the following polarity: positive, negative, and negative, positive.

During the positive half-cycle in the secondary winding, the upper diode conducts, enabling the signal to pass as shown in this diagram. The voltage developed across the load resistor becomes the output voltage. Then, during the negative half-cycle, the lower diode conducts, allowing the current or signal to flow through the load resistance R. The voltage developed across R is again as shown in this figure. Consequently, the output exhibits this type of voltage, which is referred to as full-wave rectified voltage.

This circuit effectively converts the primary AC waveform, an alternating current signal, into a pulsating DC waveform, as depicted in the output diagram. Thus, it functions as a full-wave rectifier.

c. Full wave bridge rectifier circuit | AC to DC voltage converter

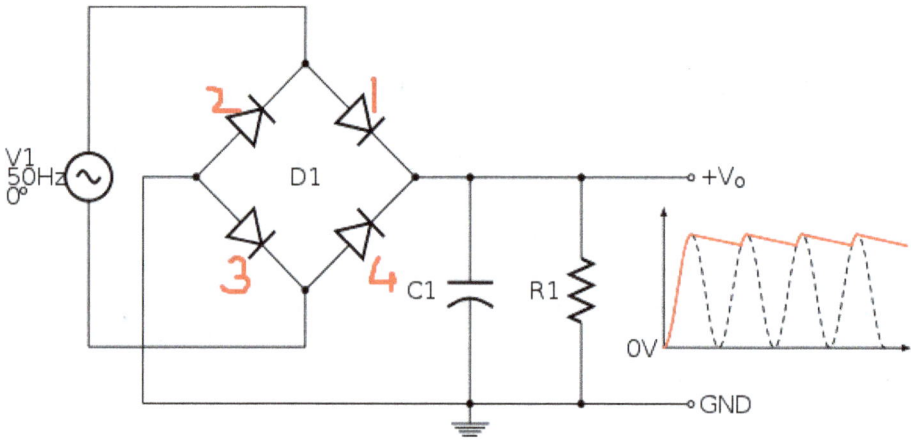

Figure 6-3 Full-wave bridge rectifier [50]

Alright, let's examine this new circuit. In this circuit, we have a sinusoidal AC voltage source with a frequency of 50 Hertz and an amplitude represented as V1. Typically, this is the AC main supply we receive in our homes, typically rated at 230 volts and 50 Hertz, delivering a sinusoidal AC power supply.

Moving on, we notice four diodes arranged in the fashion shown. Two ends of this diode circuit are connected to a sinusoidal voltage source through the provided connections. The other two points of this circuit lead to an output voltage measurement setup, as depicted here, with reference to ground. Additionally, a capacitor and a resistor are connected, and the output voltage is measured across this resistor, denoted as V_0 with respect to ground.

Now, let's delve into how this circuit operates and its application. With this sinusoidal waveform, we encounter both positive and negative cycles of oscillation, with a frequency of 50 Hertz. During the positive half-cycle, the diodes 1 and 3 becomes forward-biased, allowing the signal to flow through it and complete the circuit. This positive half-cycle voltage passes through the load resistor, labeled as R1, and is eventually grounded, forming a complete path.

This same configuration continues during the negative half-cycle when the polarity changes. This time, the diodes 2 and 4 becomes forward-biased, while the others become reverse-biased, again allowing the signal to pass through and complete the circuit. The output voltage is developed across the load resistor, with the positive and negative terminals as shown.

Now, let's look at the output waveform. The dotted black lines represent the first and second positive half-cycles. As you can see, the positive half-cycles are preserved, resulting in a pulsating DC voltage. However, it's important to note that this isn't a constant DC signal, which is typically required to be stable without any signal pulsations or variations over time.

To achieve a more constant DC signal, we introduce a capacitor. When the output voltage, due to the positive half-cycle, passes across the resistor R1, it charges the capacitor. Similarly, during the negative half-cycle, the voltage across resistor R1 again charges the capacitor, maintaining the same polarity. The capacitor's role is to charge up to the peak of the waveform and then discharge slightly before repeating the process. As a result, the red waveform represents an improved DC converted signal with a nearly constant output. This capacitor acts as a filter, removing ripples or variations from the output waveform.

However, if you were to remove the capacitor, the black shaded dotted lines represent the pulsating DC output, showing the significance of the capacitor in smoothing the signal.

In conclusion, this circuit is known as a full-wave bridge rectifier circuit, as it converts an AC waveform into a nearly constant DC waveform. To achieve the desired output, it's essential to choose suitable values for the resistor (R) and capacitor (C) to determine the charging and discharging time constants of the capacitor.

II.   Transistor circuits

In this section, we are going to explore an important BJT (Bipolar Junction Transistor) configuration. There are three key BJT configurations in which the BJT can be biased: the common emitter mode, common base mode, and common collector mode.

Let's focus on the common emitter mode of the BJT. In this circuit diagram, the input is applied to the base, the emitter is grounded, and the output is taken from the collector. Thus, the emitter serves as a common connection for both the input (base) and output (collector). The input is applied across the base-emitter junction (V_{BE}), while the output is represented as V_{CB}.

a. Transistor in common emitter mode

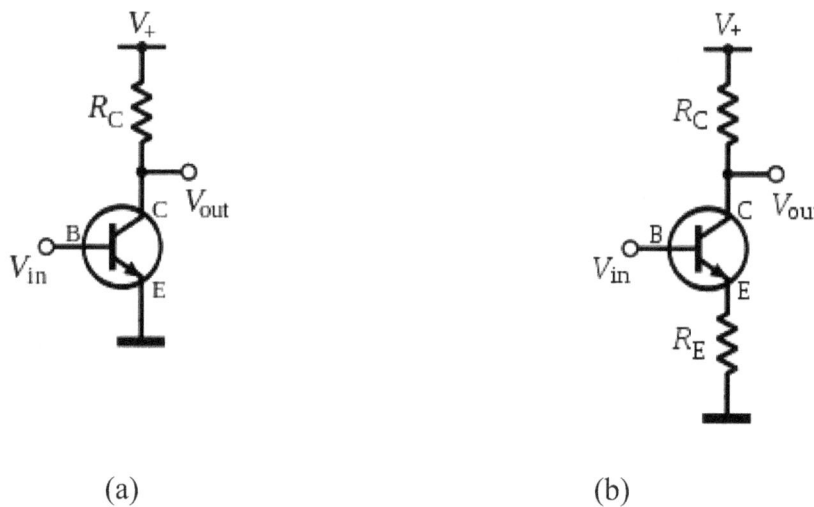

(a)   (b)

Figure 6-4 BJT in CE mode, Basic NPN common-emitter circuit (neglecting biasing details) (a), and Adding an emitter resistor decreases gain, but increases linearity and stability (b) [51]

This configuration allows the BJT to function as both a voltage and current amplifier. It typically offers a high current gain, often around 200, enabling small input currents to result in significantly larger output currents.

Additionally, it can produce voltage amplification, translating small input voltage variations into substantial output voltage changes.

However, there are specific characteristics and challenges associated with this common emitter configuration. First, it provides a medium input resistance, making it moderately sensitive to changes in input impedance. Second, it exhibits a high output resistance, which affects its ability to drive low-impedance loads effectively. Lastly, when a sinusoidal voltage is applied at the input, the output is 180 degrees out of phase with the input signal, resulting in a phase shift.

To address some of the limitations of this circuit, a modification called "emitter degeneration" can be introduced. This involves adding a small resistor (re) from the emitter to ground. This modification alters the transconductance (g_m) of the circuit, which is now expressed as $g_m = I_C / V_{BE}$. The transconductance is a measure of how the output current varies with changes in input voltage.

Emitter degeneration significantly reduces the dependence on intrinsic transistor parameters, such as g_m, and makes the circuit's gain primarily determined by the ratio of resistors, namely R_C and R_e. This modification improves gain stability and increases the input dynamic range, reducing distortion for larger input signals.

In summary, the common emitter configuration with emitter degeneration offers improved performance over the basic common emitter circuit, as it reduces gain dependency on transistor characteristics and mitigates issues

related to gain stability and input dynamic range.

b. Transistor in common base mode

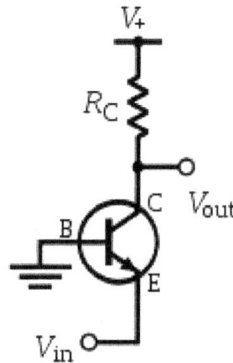

Figure 6-5 BJT in CB mode [52]

Let's take a closer look at this small circuit as shown in Figure 6-5. What we

have here is an NPN bipolar junction transistor (BJT). As we know, a BJT has

three terminals or pins: the collector, emitter, and base. It's identifiable as an

NPN transistor because the arrow on the emitter side indicates the direction of

conventional current flow, which goes from the collector to the emitter

through this transistor.

Now, in this circuit, we can observe that the base of the transistor is connected

to the ground, and the input signal V_{in} is applied to the emitter. The output is taken from the collector. Thus, we can say that the input is applied to the emitter with respect to the base, making the input known as V_{BE} (Voltage between Base and Emitter). The output is taken from the collector with respect to the base, making the output V_{CB} (Voltage between Collector and Base).

Here, the base is common between the input and output, meaning it's connected to both the emitter and the collector. This configuration is referred to as the 'common base' configuration of the transistor.

It's worth noting that you can also use a PNP transistor in the common base mode, depending on your circuit requirements. As a practice exercise, you can try drawing the PNP circuit in the common base configuration.

c. DC Biasing of the Transistor | Common Base | BJT

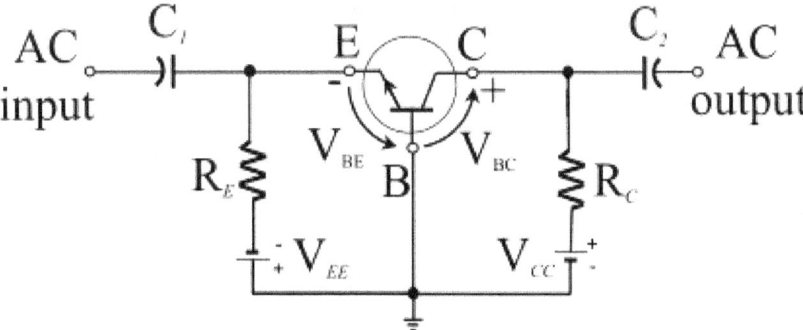

Figure 6-6 example circuit of BJT in CB mode [53]

Let's take a closer look at this circuit and try to understand its operation. This circuit takes an AC input signal and produces an AC output signal. It contains a bipolar junction transistor (BJT) configured as an NPN transistor. As indicated by the arrow on the emitter side, the conventional current flows from the collector to the emitter, classifying this transistor as an NPN type. The BJT has three terminals: collector, base, and emitter. The collector is made of an N-type semiconductor, the emitter is also N-type, and the base is P-type.

To operate this circuit as an amplifier, it needs to be biased with DC voltage sources to set its operating point or DC bias point. Let's examine the biasing and circuit components in detail:

1. AC Input: An AC input signal is applied to the circuit. Capacitor C_1 filters out any DC component, allowing only the pure AC signal to pass.

2. Emitter Biasing: To bias the transistor, a voltage source labeled V_{EE} is connected with its positive terminal to the base and the negative terminal to the emitter, through resistor R_e. This biasing configuration forward-biases the base-emitter junction, creating a voltage V_{BE} across it.

3. Output Biasing: Another voltage source labeled V_{CC} is connected with its negative terminal to the base and its positive terminal to the collector through resistor R_C. This configuration reverse-biases the base-collector junction.

The transistor operates in an active region due to the forward-biased base-emitter junction and reverse-biased base-collector junction. The biasing conditions enable the transistor to function as a common-base amplifier, with the base serving as a common terminal between the collector and the emitter. Capacitor C_2 filters out any DC component from the output signal, providing an AC output at the designated node. In summary, this circuit utilizes an NPN BJT operating in common-base mode as an amplifier. It is crucially biased to ensure proper functionality.

d. How do you bias the transistor for operation? | common base |

PNP BJT example

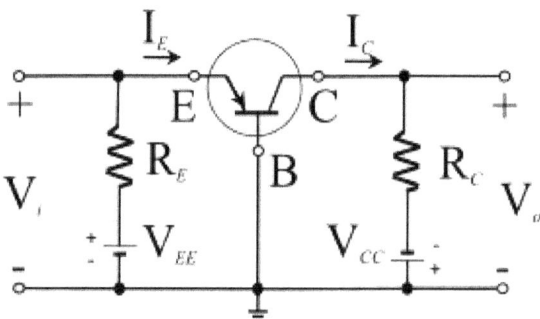

Figure 6-7 Bias the transistor in common base using PNP BJT example [54]

Let's take a closer look at another circuit example. Here, we can see a PNP

transistor connected as shown. It's a PNP transistor, characterized by its three

terminals: collector, base, and emitter. The direction of the current is from

the emitter to the collector, which is a characteristic of PNP transistors. In

PNP transistors, the majority of charge carriers responsible for the current

are holes.

Let's label the terminals: 'p' for the emitter, 'n' for the base, and 'p' for the

collector. We have an external voltage source, 'V_i,' providing an AC

alternating current signal applied across these two terminals. 'V_o' represents the output voltage, which is also an AC signal taken across these two terminals.

To provide biasing to this BJT (Bipolar Junction Transistor) circuit and set its operating point, we have two voltage sources, 'V_{EE}' and 'V_{CC},' connected via resistors, as shown. When we refer to DC biasing, we mean setting the operating point or quiescent point (Q-point) around which the externally applied AC signal will be processed, effectively amplifying it.

Let's examine the biasing conditions:

1. The positive terminal of 'V_{EE}' connects to the emitter through the resistor 'R_E,' and the negative terminal goes to the base. This forward-biases the emitter-base junction, creating a voltage 'V_{BE}' across this junction.

2. The negative terminal of 'V_{CC}' connects to the collector (P-type material), while the positive terminal goes to the base (N-type material) through the resistor 'R_C.' This arrangement reverse-biases the base-collector junction, resulting in a voltage 'VCB' across this junction.

With the emitter-base junction forward-biased and the base-collector junction reverse-biased, the transistor is biased in the active region, allowing it to function as an amplifier. This is how the circuit operates.

e. **Transistor in common collector mode | Emitter follower | voltage follower**

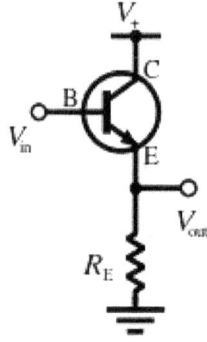

Figure 6-8 BJT in CC mode [55]

In this circuit example (Figure 6-8), we observe an NPN transistor. The direction of conventional current indicates that it flows out of the emitter. This characteristic identifies it as an NPN transistor, where the current flows from the collector to the emitter.

The three terminals of the transistor are the collector, base, and emitter. The N-region, P-region, and N-region arrangement confirms that this is an NPN transistor.

The input signal is applied to the base terminal, and the output is taken from the emitter. This configuration is known as the common collector (CC) configuration, where the collector is connected to the positive supply voltage.

The purpose of this common collector configuration is to bias the transistor at a DC operating point using external resistors (not shown). This biasing ensures that the base-to-emitter junction is forward-biased, while the base-to-collector junction is reverse-biased. Such biasing places the transistor in the active region.

When an external AC input signal is applied to the base terminal, it results in base current flowing into the base terminal. The emitter current (I_E) and collector current (I_C) flow accordingly.

Neglecting the base current contribution, we can state that I_E equals I_C. Considering Kirchhoff's voltage law (KVL) around the loop, we have V_{in} - V_{BE} - $I_E \times R_E = 0$.

By substituting $I_E \times R_E$ with V_{out}, we find that $V_{out} = V_{in} - V_{BE}$. This equation demonstrates that the output voltage in this circuit will always be equal to the input voltage minus the base-emitter junction voltage drop, typically around 0.6 to 0.7 volts.

This circuit is known as a voltage follower circuit, which is a common application of a BJT biased in common collector mode. It is also referred to as an emitter follower because the voltage at the emitter closely tracks the voltage at the base.

f. How will transistor maintain the constant bias in this circuit?

Let's revisit the circuit in Figure 6-8 in our previous module. As observed, the output (V_o) closely tracks the input (V_{in}), where the voltage drop V_{BE} is consistently maintained across the base-emitter junction.

To understand this circuit, let's consider a scenario where there is a variation in the input voltage (V_{in}), either increasing or decreasing. For instance, if V_{in} increases, the base current (I_B) of the transistor also increases due to the transistor's action. The collector current (I_C) is the product of beta (β) multiplied by I_B, where β represents the current gain of the transistor. Consequently, with increased V_{in}, I_B and I_C increase, and the collector current approximately equals the emitter current (I_E) that flows through the resistor R_E.

As a result, the voltage drop across R_E, which is equivalent to V_{out}, increases. Therefore, when V_{in} rises, V_{out} endeavors to catch up, and this relationship is proportional.

Conversely, when V_{in} decreases, I_B and I_C decrease, causing a reduction in I_E, which in turn leads to a decrease in the voltage drop across R_E, resulting in a lower V_{out}. Hence, for both increases and decreases in V_{in}, V_{out} adapts proportionately.

This phenomenon ensures that the voltage drop V_{BE} remains constant. The voltage at the base is at the positive terminal, while the emitter voltage is at the negative terminal. This constant V_{BE} drop is maintained, allowing the transistor to maintain equilibrium in the circuit.

It's worth noting that variations in the base-emitter voltage (V_{BE}) may occur due to changes in the DC biasing. In such cases, the transistor responds by adjusting the voltage at the emitter accordingly, ensuring that the V_{BE} drop is consistently maintained. This circuit configuration is known as the common collector configuration, where the output closely follows the input.

g. Fixed bias or base bias - 1

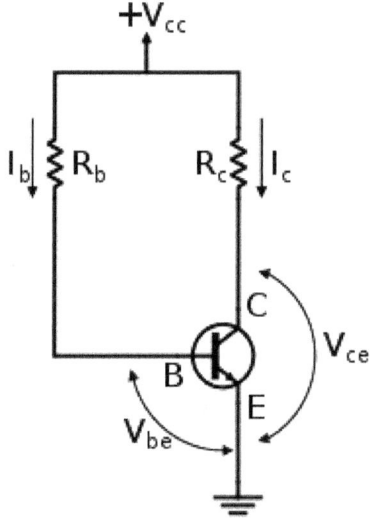

Figure 6-9 Fixed bias circuit (Base bias) [56]

In this circuit (Figure 6-9), we're examining one of the fundamental biasing techniques for Bipolar Junction Transistors (BJTs). This configuration is known as the fixed bias or base bias circuit, often referred to as base bias because it uses fixed resistor values (R_B and R_C) to establish the DC operating point or Q point of the circuit. Let's delve into how it operates.

In this setup, we have an NPN transistor with three terminals: collector, base, and emitter. The voltage across the base and emitter is termed V_{BE}, while the voltage across the collector and emitter is denoted as V_{CE}.

For the input side, KVL yields: $V_{CC} - I_B R_B - V_{BE} = 0$

For the output side, KVL yields: $V_{CC} - I_C R_C - V_{CE} = 0$

From these equations, we can compute I_B, which is related to I_C through the transistor's current gain (β) with the equation $I_C = \beta * I_B$. This base current, I_B, enables us to determine I_C and V_{CE}.

It's worth noting that in this configuration, the base-emitter junction is forward-biased, while the collector-base junction is reverse-biased, placing the transistor in the active region of operation.

Now, let's discuss the drawbacks of this circuit. Firstly, as the temperature increases, causing the collector current (I_C) to rise, the base current (I_B) also increases due to the transistor equation. This shift in I_C and I_B alters the original operating point, impacting circuit stability.

Secondly, variations in the transistor's current gain (β) can result in different operating points, further affecting circuit stability. To mitigate these stability issues related to temperature and β variations, modifications like adding an emitter resistor (R_E) in the emitter path are often employed.

h. Fixed bias or base bias - 2

Let's take a closer look at this Bipolar Junction Transistor (BJT) circuit. This circuit is known as fixed bias with emitter resistor or fixed bias with emitter degeneration resistor. The emitter resistor is positioned here to enhance circuit stability and reduce harmonic distortion.

As you can observe, it features an NPN transistor with collector, base, and emitter pins. The base-emitter voltage is denoted as V_{BE}, and the collector-emitter voltage is V_{CE}. The term "fixed bias" originates from the fact that we apply a supply voltage, V_{CC}, and connect a resistor to the base terminal, which establishes biasing using the supply voltage itself.

In this configuration, two Kirchhoff's voltage law equations can be applied:

1. $V_{CC} - I_B R_B - V_{BE} - I_E R_E = 0$ Here, I_B represents the current through the base resistor, V_{BE} is the voltage across the base-emitter junction, and $I_E R_E$ signifies the emitter current.

2. $V_{CC} - I_C R_C - V_{CE} - I R_B - I_E R_E = 0$ This equation accounts for the current through the collector resistor (R_C), the collector-emitter voltage (V_{CE}), the current through the base resistor (R_B), and the emitter resistor (R_E).

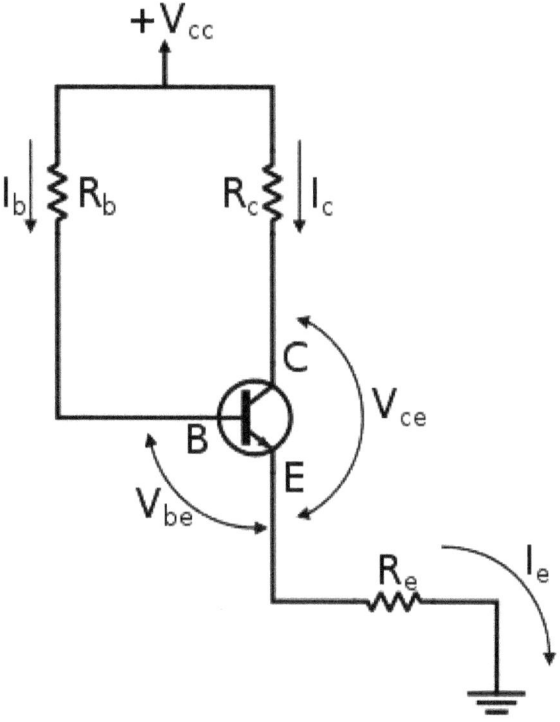

Figure 6-10 Fixed bias with emitter resistor [56]

This biasing arrangement is termed "fixed bias" because biasing is achieved through a fixed resistor, R_B, through which the base current flows. The fundamental operation of this NPN BJT transistor remains consistent in this configuration.

i. collector base bias or collector feedback biasing

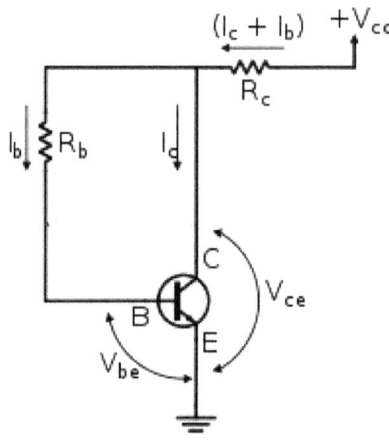

Figure 6-11 Collector-to-base bias [56]

This is another circuit used for biasing the BJT transistor, known as the collector-to-base bias circuit (Figure 6-11). In this configuration, a supply voltage V_{CC} is applied, and we have resistors R_C and R_B connected to the circuit. This arrangement earns it the name "collector-to-base register biasing" for the BJT.

As with most circuits, this one involves an NPN transistor with collector, emitter, and base pins. The base-emitter junction is forward-biased, while the base-collector junction is reverse-biased, setting the transistor's operating point in the active region. Resistors R_B and R_C are employed to establish the circuit's operating point.

When the supply voltage is activated, current flows into the circuit, denoted as "I." This current splits into two branches, one being I_C and the other I_B. So, essentially, the total current I is the sum of I_C and I_B.

Applying Kirchhoff's Voltage Law (KVL) to the circuit, we have V_{CC} - voltage drop across R_C - voltage drop across R_B - V_{BE} = 0. Utilizing this KVL equation allows us to calculate current and voltage values within this circuit loop.

R_B and R_C play a crucial role in setting the circuit's operating point. There are two primary factors that can change the operating point, and we will examine how this circuit helps stabilize it.

Firstly, changes in temperature can alter the operating point. For a fixed V_{BE}, if the circuit's temperature increases, the collector current starts to rise. Since collector current is related to base current through the equation $I_C = \beta * I_B$, an increase in I_C leads to an increase in I_B. This, in turn, changes the operating point due to temperature rise.

The circuit responds to the temperature rise by causing the voltage drop across R_C to increase because of the increasing I_C. Consequently, the voltage drop across R_B decreases. According to KVL, V_{RB} (voltage across R_B)

decreases, causing I_B to decrease. This phenomenon, termed negative feedback, counteracts the initial increase in I_B due to rising temperature.

Secondly, transistor beta (β) variations can also affect the operating point. Replacing the transistor in the circuit with one having a different β will result in a different I_C and I_B. However, by connecting resistor R_B from the collector to the base, the circuit mitigates the impact of β variations on the DC operating point, providing stability in the face of changes in temperature or transistor beta.

In summary, this collector-to-base bias circuit is designed to stabilize the operating point against variations in temperature and transistor beta, ensuring consistent performance under different conditions.

j. Voltage divider bias circuit for transistor | BJT biasing

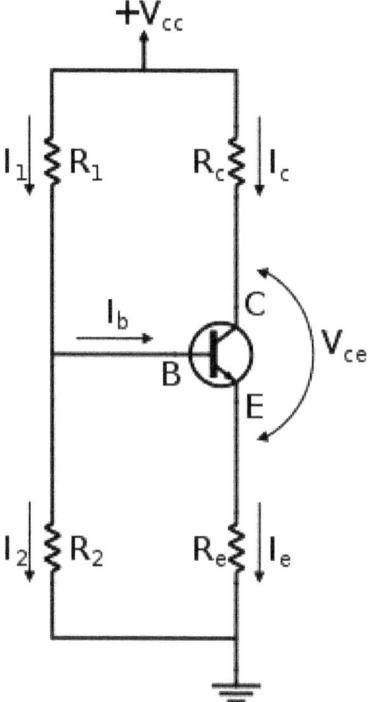

Figure 6-12 BJT voltage divider bias [56]

Here is one of the most popular biasing techniques for bipolar junction transistors, known as the voltage divider bias circuit (Figure 6-12). In this configuration, we have an NPN transistor with three pins: collector, base, and emitter. The voltage across the base and emitter is referred to as V_{BE}, and the voltage across the collector and emitter is denoted as V_{CE}.

To set the DC operating point for this transistor, we use two resistors, R1 and R2, along with a supply voltage, V_{CC}. The voltage division created by R_1 and R_2 applies a voltage across the base-emitter junction, which can be calculated as follows:

$V_{R2} = V_{CC} * (R_2 / (R_1 + R_2))$

This is a simple application of the potential divider rule. The resulting voltage, V_{R2}, is applied across the base-emitter junction. Next, there is an emitter current (I_E) that flows through the resistor R_E. To analyze this circuit, we can apply Kirchhoff's voltage law (KVL) to the loop involving V_{R2}, V_{BE}, and I_E.

Similarly, for the outer circuit of the BJT, we can apply KVL to the loop involving V_{CC}, $I_C R_C$ (collector current through RC), and V_{CE}. These two loops help us determine the voltages and currents in the circuit.

The base current (I_B) can be calculated as the difference between I_1 and I_2:

$I_B = I_1 - I_2$

This voltage division circuit allows us to establish a DC operating point for the transistor.

Please note that the specific values of currents and voltages would depend on the component values and transistor characteristics used in the circuit.

Chapter 7 Principles of electronic

Measurements

I. Measuring voltage, current, and resistance

a. How to measure resistance using digital multi-meter (DMM)?

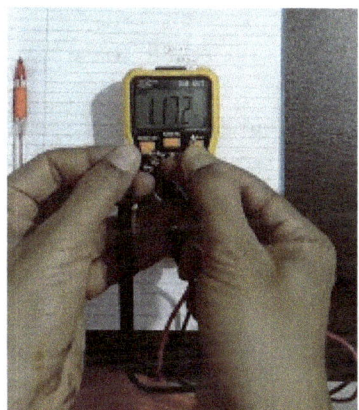

Figure 7-1 Resistance measurement using DMM

Measuring resistance using a Digital Multi meter (DMM) is a fundamental and straightforward process commonly used in electronics and electrical troubleshooting. Here are the steps to measure resistance using a DMM:

Steps to Measure Resistance with a DMM:

1. **Turn On the DMM:**

- Power on the DMM by rotating the mode dial or pressing the power button, depending on the model.

2. **Select Resistance (Ω) Mode:**

- Set the DMM to the resistance (Ω) measurement mode. This mode is usually indicated by the Greek letter omega symbol (Ω) on the DMM's mode dial.

3. **Choose the Range:**

- If your DMM offers multiple resistance measurement ranges, select a range that is equal to or greater than the expected resistance of the component you are testing. It's best to start with the highest range and work your way down for accuracy.

4. **Connect the Test Leads:**

- Connect the black test lead (negative or common) to the common (COM) terminal on the DMM.

- Connect the red test lead (positive) to the terminal labeled for resistance measurements (Ω or resistance).

5. **Measure the Resistance:**

- Place the test leads across the component or resistor you want to measure. For through-hole resistors, touch the leads to each resistor terminal. For SMD resistors, use tweezer probes or fine-point probes to make contact with the resistor's terminations.

6. **Read the Measurement:**

   - The DMM will display the resistance value on its screen. If you're using a digital multimeter, it will show the value in ohms (Ω). If you're using an analog multimeter, read the resistance value from the scale and consider any multiplier markings (e.g., $k\Omega$ for kilohms).

7. **Record the Value:**

   - Note down the measured resistance value for reference.

8. **Disconnect the Test Leads:**

   - Remove the test leads from the component you've measured.

9. **Turn Off the DMM:**

   - To conserve battery life, turn off the DMM when you've completed your measurements.

Chapter 8   Additional topics

I.   Integrated circuits (ICs)

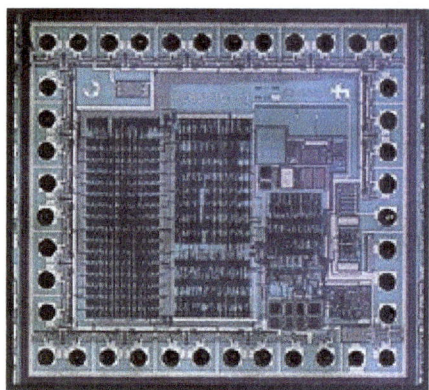

Figure 8-1 A microscope image of an integrated circuit (IC) die. The pinouts are the black circles surrounding the IC [57].

Figure 8-1 shows us a micrograph of an IC, which stands for an integrated circuit. An IC stands for "integrated circuit." This micrograph depicts a piece of silicon. Integrated circuits are typically made from semiconductor material, such as silicon, which is clearly visible here. This is also commonly referred to as a "chip," "die," or "IC die." When examining the details and microfeatures of the electronic components integrated onto this die, a microscope is required.

Usually, the dimensions of these microchips are in the range of just one or two square millimeters. For example, the height and width of such a chip can be as small as 1 mm x 1 mm.

The central area of this chip is where various electronic components, devices, and circuits have been integrated. These chips can serve various purposes, such as microprocessors, microcontrollers, memory, analog circuits, radio frequency circuits, and other digital logic components, all compactly integrated within this small area.

The black circles located around the perimeter of the IC represent the pinout. They indicate the connection points for the circuits within. In the modern era of microelectronics, electronic components have been miniaturized and integrated onto pieces of silicon, leading to a significant reduction in the cost of electronics. Microprocessors or microcontroller ICs, for instance, are now available at very affordable prices, making them accessible to a broader audience.

The development, manufacturing, and testing of integrated circuits require a range of skills and expertise. Designing, producing, and ensuring the quality of these ICs are essential steps in the electronics industry.

II.   Analog and Digital | electronic | basics and advance concepts

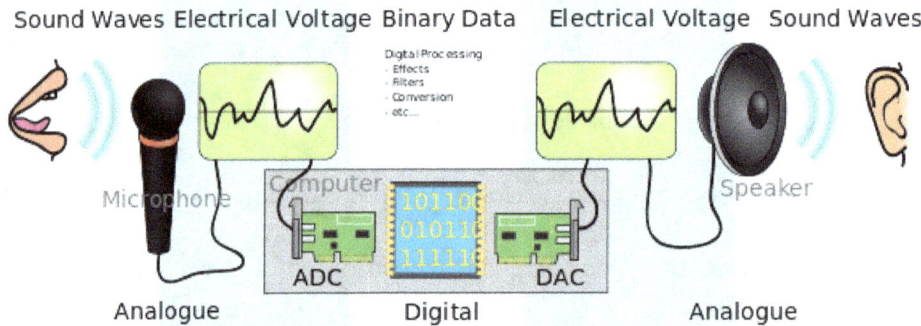

Figure 8-2 Analogue and digital world [58]

In Figure 8-2, we have a detailed illustration explaining a fundamental concept in electronics – the transmission and reception of electronic signals, particularly in the domain of acoustics or sound. This serves as an excellent opportunity to grasp the difference between analog and digital signals and understand how these signals are processed, including the instruments and devices used to convert one form of energy into another. Central to this process are transducers or sensors.

To the far left in the image, you'll notice someone speaking. This spoken audio is represented by sound waves, which can be considered acoustic signals. These signals possess specific amplitudes and frequencies, typically ranging from 20 Hertz to 20 kilohertz. To process these sound waves or acoustic signals, we employ an electronic device known as a microphone, which acts as a transducer. It captures the sound vibrations and converts them into proportional electrical voltages. This transformation is depicted in the accompanying graph, demonstrating an analog signal with electrical voltage that continuously varies with respect to time. It exhibits both positive and negative amplitudes, creating an instantaneous signal concerning time.

Now, turning our attention to the rectangular box, we find a computer. Within the computer, essential components include an Analog-to-Digital Converter (ADC), a Digital-to-Analog Converter (DAC), and digital circuits. The ADC primarily converts the analog voltage, as seen in the graph, into digital codes, effectively transitioning from the analog domain to the digital domain. These digital codes are then subjected to further processing using Digital Signal Processing (DSP) circuitry, which may involve various operations like filtering, amplification, and modulation.

The digital codes generated during this processing phase are then sent to the Digital-to-Analog Converter (DAC), which performs the reverse transformation, converting digital codes back into proportional electrical voltages. In this manner, the signal transitions from the digital domain back to the analog domain, and we have an analog signal, as portrayed in the second graph.

Finally, this analog signal is fed to another transducer, a loudspeaker, which converts the incoming electrical signal into sound waves proportional to the original acoustic signal. These sound waves are then presented to the human ear for audible reception.

In this manner, the complete communication system operates – from the source (such as a speaker) to the receiver (the listener). Electronics plays a pivotal role in enabling this transmission of sound and information. We hope you found this explanatory topic informative and appreciate the role of technology in transducing analog and digital signals.

III.   **What is noise in Electronics? What causes it?**

In this module, we take a closer look at Figure 8-3 to understand the concept of noise in typical electronic systems and electronic signals. In electronics, there are two types of signals: analog and digital. Let's start by understanding what an analog signal is. On the y-axis, the signal can represent current, voltage, or a combination of both, and it continuously varies with respect to time. On the other hand, a digital signal switches between two levels, often referred to as high and low levels or positive and negative levels. These two logic levels define digital signals.

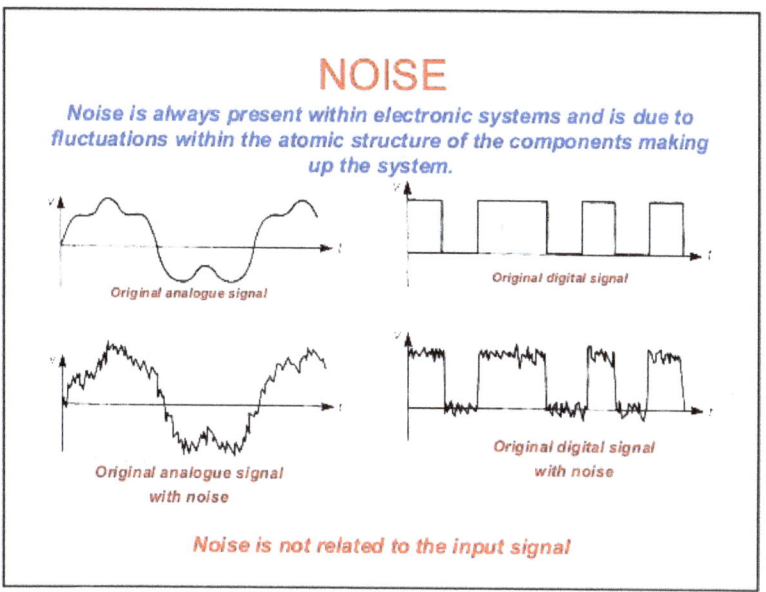

Figure 8-3 Electronic Noise [59]

Now, let's get back to our main topic, which is noise. Whether you have an analog or digital signal, noise is always present in electronic systems. The components and devices used to process these signals, whether analog or digital, produce fluctuations due to the atomic structures of these components. These atomic fluctuations are introduced into the system.

When noise is added to the original signals, the signal may look like this. In the case of an analog signal, the original signal gets mixed with noise and may appear corrupted. Similarly, with a digital signal, the original signal may look like this when noise is present.

In practical terms, think of it as an example: when I'm explaining this concept to you through my voice, my original signal, which is my voice, is reaching you. However, in the background, there's additional audio, which we can refer to as noise, mixing with my voice. This illustrates how original signals can get corrupted when noise is introduced into the system.

Hope you liked the contents of this book, if you did so, do share with others and visit our website for business enquiries and communication.

Visit our YouTube channel and subscribe it for updates and valuable contents.

Wish You Happy Learning!

References

[1] "No Title." https://en.wikipedia.org/wiki/Electronics#/media/File:Componentes.JPG

[2] "No Title." https://commons.wikimedia.org/wiki/File:Diod_LED_symbol.png

[3] "No Title", [Online]. Available: https://commons.wikimedia.org/wiki/File:Carbon_Composition_Resistor_1K3_cracked.png

[4] "No Title", [Online]. Available: https://commons.wikimedia.org/wiki/File:10k_Ohm_Potentiometer_-_POT-103-BIG_%284149943231%29.jpg

[5] "No Title", [Online]. Available: https://commons.wikimedia.org/wiki/File:10k_Ohm_Breadboard_Compatible_Potentiometer.jpg

[6] "No Title", [Online]. Available: https://commons.wikimedia.org/wiki/File:100_ohm_SMD_1206_resisto

r.jpg

[7]   "No   Title",   [Online].   Available:

https://commons.wikimedia.org/wiki/File:22_ohm_SMD_0603_resistor

.jpg

[8]   "No   Title",   [Online].   Available:

https://www.digikey.in/en/resources/conversion-calculators/conversion-

calculator-smd-resistor-code

[9]   "webpage."   https://www.utmel.com/tools/smd-resistor-code-

calculator?id=33

[10]   "No   Title."

https://commons.wikimedia.org/wiki/File:VFPt_capacitor_mod.svg

[11]   "No   Title."

https://commons.wikimedia.org/wiki/File:Electronic_component_capac

itors.jpg

[12]   "No Title." https://creativecommons.org/licenses/by-sa/4.0/

[13]   "No Title." https://creativecommons.org/licenses/by-sa/2.0/

[14]   "No Title." https://creativecommons.org/licenses/by/2.0/

[15]   "No Title." https://creativecommons.org/licenses/by-sa/3.0/

[16] "No Title." https://commons.wikimedia.org/wiki/File:Inductor_Symbols.jpg

[17] "No Title." https://en.wikipedia.org/wiki/Alternating_current

[18] "No Title." https://commons.wikimedia.org/wiki/File:Conventional_Current.png

[19] "No Title", [Online]. Available: https://en.wikipedia.org/wiki/File:Non-ideal_voltage_and_current_sources.svg

[20] "No Title", [Online]. Available: https://commons.wikimedia.org/wiki/File:Block_resistance.svg

[21] "No Title", [Online]. Available: https://commons.wikimedia.org/wiki/File:Ohm%27s_law_triangle_%28 VIR%29.jpg

[22] "No Title." https://en.wikipedia.org/wiki/Kirchhoff's_circuit_laws

[23] "No Title." https://et.wikipedia.org/wiki/Kirchhoffi_seadused

[24] "No Title." https://commons.wikimedia.org/wiki/File:Kirchhoff%27s_current_law.jpg

[25] "No Title." https://en.wikipedia.org/wiki/Series_and_parallel_circuits

[26] "No   Title."

https://bn.wikipedia.org/wiki/রোধক#/media/চিত্র:Series_resistors.jp

g

[27] "No   Title."

https://su.wikipedia.org/wiki/Résistor#/media/Gambar:Resistorscombo.

png

[28] "No   Title."

https://en.wikipedia.org/wiki/Reflections_of_signals_on_conducting_li

nes

[29] "No Title." https://commons.wikimedia.org/wiki/File:Voltage_divider-

loaded.svg

[30] "No Title." https://en.wikipedia.org/wiki/Current_divider

[31] "No   Title."

https://commons.m.wikimedia.org/wiki/File:Energy_bands_of_a_semic

onductor_type_N.svg

[32] "No Title." https://commons.wikimedia.org/wiki/File:ಅರೆವಾಹಕ.png

[33] "No   Title."   https://commons.wikimedia.org/wiki/File:SY320-

10_HFO.jpg

[34] "No   Title."
https://commons.wikimedia.org/wiki/File:Germanium_Diode_1N60P.jp
g

[35] "No   Title."   https://commons.m.wikimedia.org/wiki/File:V-
a_characteristic_diodes_si_ge.svg

[36] "No Title."

[37] "No   Title."
https://commons.m.wikimedia.org/wiki/File:Transistors.agr.jpg

[38] "No   Title."
https://commons.wikimedia.org/wiki/File:NPN_BJT_Basic_Operation_
%28Active%29_fr.svg

[39] "No   Title."
https://commons.wikimedia.org/wiki/File:NPN_AND_PNP_BJT_SYM
BOLS.png

[40] "No   Title."
https://upload.wikimedia.org/wikipedia/commons/9/91/Transistor_Simp
le_Circuit_Diagram_with_NPN_Labels.svg

[41] "No Title." https://en.wikipedia.org/wiki/Field-effect_transistor

[42]   "No Title." https://et.wikipedia.org/wiki/Pn-siirdega_väljatransistor

[43]   "No   Title."
https://en.wikipedia.org/wiki/List_of_MOSFET_applications

[44]   "No   Title."
https://commons.wikimedia.org/wiki/File:FET_Symbols.svg

[45]   "No Title." https://en.wikipedia.org/wiki/File:MOSFET_transistors.jpg

[46]   "No   Title."
https://en.wikipedia.org/wiki/File:Scheme_of_metal_oxide_semiconduc
tor_field-effect_transistor.svg

[47]   "No   Title."
https://en.wikipedia.org/wiki/CMOS#/media/File:Cmos_impurity_profi
le.PNG

[48]   "No   Title."
https://commons.wikimedia.org/wiki/File:SE_Half_Wave_Rectifier.svg

[49]   "No   Title."
https://commons.wikimedia.org/wiki/File:SE_Full_wave_rectifier_-
_Center_tapped.svg

[50]   "No Title." https://commons.wikimedia.org/wiki/File:RC_filter.svg

[51] "No Title." https://en.wikipedia.org/wiki/Common_emitter

[52] "No Title."

[53] "No   Title."
https://commons.m.wikimedia.org/wiki/File:NPN_BJT_common_base_
biasing-en.svg

[54] "No   Title."
https://commons.m.wikimedia.org/wiki/File:PNP_BJT_common_Base_
bias.svg

[55] "No Title." https://en.m.wikipedia.org/wiki/Common_collector

[56] "No Title." https://en.wikipedia.org/wiki/Bipolar_transistor_biasing

[57] "No Title." https://en.wikipedia.org/wiki/Integrated_circuit

[58] "No   Title."   https://en.wikibooks.org/wiki/A-
level_Computing/AQA/Paper_2/Fundamentals_of_data_representation/
Analogue_and_digital

[59] "No   Title."   https://commons.m.wikimedia.org/wiki/File:Electronics-
noise-1-638.jpg

[60] "No Title." https://en.wikipedia.org/wiki/Electronics